KAMPANGA, A FIVE-YEAR-OLD FEMALE MOUNTAIN GORILLA IN THE VIRUNGA MOUNTAINS OF AFRICA

To Gail

Nick

THE GREAT APES

Between Two Worlds

Photographs and essays
by Michael Nichols

Contributions by
Jane Goodall
George B. Schaller
Mary G. Smith

Prepared by the Book Division
National Geographic Society,
Washington, D.C.

PAGES 2-3: Caged alone for years in an African zoo, an old male chimpanzee reaches out to a rare understanding stranger: primatologist Jane Goodall, who calmed his outburst of rage by bowing in a chimp's pose of submission.

HARDCOVER STAMP: Image taken from a fine first-edition copy of Paul Du Chaillu's book Explorations and Adventures in Equatorial Africa *(London, 1861), presented to Dian Fossey by naturalist David Attenborough.*

ENDSHEETS: An African equatorial forest—shown here in a monochrome rendering—represents the type of habitat required by the great apes.

THE GREAT APES
Between Two Worlds

Jane Goodall
Michael Nichols
George B. Schaller
Mary G. Smith
Authors

Michael Nichols
Photographer

Published by
**The National
Geographic Society**

Gilbert M. Grosvenor
President and
Chairman of the Board

Michela A. English
Senior Vice President

Prepared by
The Book Division

William R. Gray
Vice President
and Director

Margery G. Dunn
Charles Kogod
Assistant Directors

Staff for this Book

Charles Kogod
Project Editor
and Illustrations Editor

Mary Ann Harrell
Text Editor

Cinda Rose
Art Director

Victoria Cooper
Bonnie S. Lawrence
Researchers

Jody Bolt
Seymour L. Fishbein
Edward Lanouette
Jennifer C. Urquhart
Picture Legend Writers

Carl Mehler
Map Research

Sandra F. Lotterman
Editorial Assistant

Artemis S. Lampathakis
Illustrations Assistant

Heather Guwang
Production Project Manager

Lewis R. Bassford
H. Robert Morrison
Richard S. Wain
Production

Elizabeth G. Jevons
Teresita Cóquia Sison
Marilyn J. Williams
Staff Assistants

Neva L. Folk
Special Assistant

Dianne L. Hardy
Indexer

Manufacturing and
Quality Management

George V. White
Director

John T. Dunn
Associate Director

Vincent P. Ryan
Manager

R. Gary Colbert
Executive Assistant

Contents

Cuddled by her mother, Papoose, baby Pasika embodies utter dependence—and hope for her kind. Only about 600

mountain gorillas survive in Uganda and the Virunga region of Zaire and Rwanda.

In a national park in Zaire, tourists come dangerously close to a big silverback gorilla—an adult male thoroughly

accustomed to visitors, but at risk from infectious human diseases.

Foundlings like these two-year-old lowland gorillas find foster care at the Brazzaville Gorilla Orphanage in Congo.

Hunters out for "bush meat" shoot the mothers and try to sell the young.

Wild-born chimp turned artificial human, with front teeth removed for safety reasons, 28-year-old Mr. Jiggs entertains

a New Jersey party, roller-skating among guests and kissing their hands. 13

A Prologue

Creatures of legend and anecdote for centuries, the great apes of the wild have become known to science in just three decades. But even as their lives are revealed, they face extinction. Field researcher turned crusader, Jane Goodall campaigns to save them. She illustrates one of her lectures in New York City with a portrait of a baby chimpanzee called Galahad (at left).

By Jane Goodall

15

Deep in the forest, I sat near the small group of chimpanzees. Fifi groomed her daughter Fanni. Fifi's son Faustino, three and a half years old, stared intently at the newborn infant she cradled—his brother Ferdinand. Fanni's infant, Fax, reached out to pull one of Ferdinand's small hands, but Fifi pushed her grandson away. Nearby, Fanni's young sister Flossi played by herself, laughing as she stimulated that ticklish place between neck and shoulder with a rough stone. Fifi's two oldest sons, Freud and Frodo, were away with the other adult males, courting a female.

My mind slipped back easily to a time 21 years before, when I had sat thus with Fifi's much-loved mother, the old matriarch Flo. That was in 1971, when Fifi had just given birth to Freud, and her brother Flint was still alive. Flo had been dead now for twenty years, and Flint too, for he had died grieving for his mother. Flo's two eldest sons had been alive then: Faben, who lost the use of one arm in the polio outbreak of 1966 and learned to walk long distances upright, and Figan, who with Faben's support had risen to top position in the male hierarchy and reigned for eight years. Eventful years when we learned that chimpanzees, like humans, have a dark side to their nature: They are aggressively territorial, and may hunt down members of neighboring communities and assault them fatally. We even observed a few cases of infanticide and cannibalism.

Fifi, the prolific male Evered, and the sterile female Gigi are the only survivors of the chimps I knew when I arrived at Gombe in 1960 to begin what has become the longest field study of any animal group in the wild. In the sixties, those of us working with the great apes in the field all knew one another—we were a tiny band. Today, several hundred people from around the world are actively engaged in field studies, and many more are working on ape behavior in captivity. And the nature of fieldwork has changed dramatically. In those days everything was new; almost nothing was known about the world of our closest relatives. We watched, we wrote in our little field notebooks, and we marveled at the complex behavior that we observed, the fascinating characters that we gradually came to know. Today, students come into the field with specific questions and make detailed studies of particular aspects of behavior.

I began on my own at Gombe, but built up an interdisciplinary team of students from around the world. Then, in 1975, four students were kidnapped by rebels from Zaire and held for ransom. Eventually they were returned, after an incredibly harrowing experience. After that the Tanzanian field staff, already well trained, took over the research. Even now they make most of the day-to-day observations. These men lack formal higher education, but they understand the chimpanzees and they excel in following them around over rugged terrain. They make detailed written reports, use state-of-the-art tape recorders, and operate video cameras charged from solar panels.

During my last visit Yahaya Alamasi, a born cameraman, showed

Skilled by years of practice and by the example of her own mother, Fifi tends baby Faustino, youngest of five offspring in 1989. "Chimpanzee infants seem to have unlimited energy," remarks Jane Goodall. Often Fifi romps with her children—and sometimes she prefers to relax and watch. Mature males normally tolerate infants, however pesky, and on occasion an adult may take a special interest in a certain youngster; but in this promiscuous species a father as such has no social role in the wild.

me one of his tapes—a tense interaction between Wilkie, the current alpha or top-ranking male, and two of his rivals, Prof and Beethoven. Goblin, the deposed alpha, appeared to be siding with Wilkie. "We thought Goblin would try to overthrow Wilkie again," Yahaya told me, "but now he seems to have given up." (He said this in his soft, musical Kiswahili, the language that is spoken across East Africa.) Hilali Matama, headman, told me about a patrol when a group of our males were chased for more than a kilometer by a larger group of neighboring males. "Goblin and the old man Evered led the patrol," he said; "Wilkie was first to turn and run away."

Our local staff has proved invaluable for conservation. Gombe is a tiny national park: 20 square miles, or 52 square kilometers. In 1960 chimpanzee habitat stretched as far as I could see. Today cultivated fields and dwellings have crept right up to the boundaries so that the chimpanzees are imprisoned as if they were on an island. And, as human populations increase and multiply, this is happening across the species' range in Africa. It is the same for the bonobos, or pygmy chimpanzees, and the gorillas. And it may become even worse for the orangutans in Indonesia.

Everywhere the homelands of our closest relatives are shrinking and the populations are increasingly fragmented, the gene pools so reduced in some places that long-term survival is virtually impossible.

At Gombe, however, although the small size of the population—150 at most—threatens its future, the chimpanzees live in a sort of paradise contrasted with those in so many other regions. And this, to a large extent, is because the field staff care about them as individuals, just as I do. Moreover, they are proud of their work and tell family and friends about it. There is no poaching at Gombe today.

In some African countries, chimpanzee meat is a delicacy. Even if mothers are not killed for food, they may be shot so that their babies can be seized and sold, either to someone who wants a "pet" or to a dealer with links to the entertainment or biomedical research industries.

In Gabon, Caroline Tutin, a former Gombe student, could not get good observations for many years, so fearful were the chimpanzees. Richard Wrangham, another Gombe alumnus now working in Uganda's Kibale Forest, told me how many of these chimpanzees get caught in snares—they suffer agonies until the hand or foot drops off, leaving the victim crippled for life, or till infection kills them. The chimpanzees in the Taï Forest of Côte d'Ivoire, studied by Christophe Boesch, also get ensnared—but some have learned to remove the wire. One adult male is thought to have freed the wrist of an adult female.

Variations of behavior, from one population or community to another, are especially fascinating; but our understanding can never be complete. Already whole populations have gone, whole cultures perished. For it is now clear that chimpanzees, like humans, have distinct cultures—

behaviors passed from one generation to the next through observation and learning. One ape's innovation can be incorporated gradually into the repertoire of the group.

In 1986 there was a researchers' conference in Chicago, "Understanding Chimpanzees." We already knew that chimpanzees in different areas use different objects as tools for different purposes. Now we learned more details. A film from Christophe Boesch showed the use of rock or tree-root anvils and stone or branch hammers to open hard-shelled nuts. No one has reported this in East Africa. Toshisada Nishida showed a film about chimpanzees in Tanzania's Mahale Mountains, some ninety miles south of Gombe; these animals extract carpenter ants from tree-trunk nests. They use twigs or stems much as chimps at Gombe use them to fish for termites. There are plenty of carpenter ants at Gombe, but they are not part of the chimps' normal diet.

Among many other intriguing differences, we recognized one constant. As one researcher after another described the situation in his or her area, it became apparent that chimpanzees, once common across Africa, were vanishing at an alarming rate. Concerned organizations petitioned the U.S. Fish and Wildlife Service, asking that chimpanzees and bonobos be reclassified from "threatened" to "endangered" species. Bonobos everywhere, and all chimpanzees in their historic African range, wild or captive, are now listed as endangered, but the status of captive chimps in other parts of the world was left virtually unchanged.

Soon afterward I saw a videotape, taken secretly in one of the medical research labs in the United States. It held infant chimpanzees crammed into inadequate little cages—two infants per cage, in cages stacked one above the other in a double tier. U.S. law requires a cage floor at least three times the size of the animal standing on all fours.

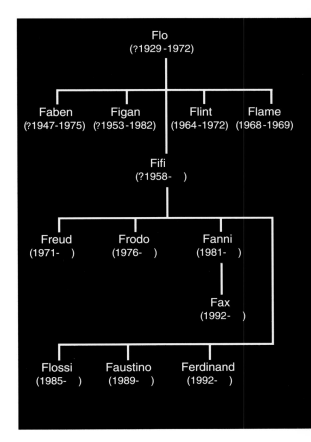

It was time, I decided, that I tried to pay back some of the debt I owe to chimpanzees. Spending less and less time in the field, I began to visit laboratories, to see conditions with my own eyes and to discuss ways of improving things with directors and technicians. And there are still zoos exhibiting chimpanzees in small, barren, utterly inappropriate cages—even in wealthy countries. JGI (the Jane Goodall Institute for Wildlife Research, Education and Conservation, a nonprofit organization) put more effort into its new program, ChimpanZoo, to improve conditions.

In recent months the institute has become deeply involved in a problem that exists in almost all the countries where great apes remain. It is the same for chimpanzees, gorillas, and orangutans: What to do with orphans whose mothers have been shot by hunters? Or with former pets now unmanageable? Ironically, the stricter the laws on international trading, the harder the problem becomes. True, there may be a local zoo, but only too often the animals are in pitiful condition—hardly surprising, considering that in many of these countries town-dwellers may be lucky if they can afford to buy one meal a day.

Sanctuaries, large enclosed areas of natural habitat, are an alternative. Baboon Island in The Gambia and Chimfunshi in Zambia already existed for chimps. Our planned sanctuary in Burundi is not yet built, but we have 17 orphans waiting in what we call the "halfway house." Near Pointe Noire in Congo, a subsidiary of the oil company Conoco Inc. has built a sanctuary for youngsters confiscated by government officials. Sadly, plans for a sanctuary in Zaire and attempts to gather orphans in Angola have been frustrated by political turmoil.

How fortunate I was that Fate set my own study in Tanzania, one of the most stable countries in Africa and one with a long history of good conservation practice. For the thirtieth anniversary of my arrival we planned a major celebration, a Wildlife Awareness Week. It was a huge success, with walks for wildlife, film shows, lectures, fund-raising dinners. I gave many talks at schools. The focus of press and radio coverage was an exhibit, "Understanding Chimpanzees," opened by the former President Julius Nyerere, "Father of the Nation." Along with photographs and videos, the exhibit displayed actual objects used as tools by chimpanzees in different parts of Africa. Some captive chimps love to paint, and a selection of paintings by different "artists" aroused great interest.

The whole week was so successful that I thought we should try to replicate it elsewhere; and Conoco and the World Wide Fund for Nature/Switzerland agreed to sponsor the venture. To date we have held wildlife weeks in Uganda, Burundi, Congo, Angola, and Sierra Leone (as well as two more in Tanzania). These succeeded because many people worked hard to organize them—and thus an ever greater network of support is being generated across the continent, linking heads of state and conservation workers and hundreds and hundreds of schoolchildren.

Some moments stand out. In Uganda, Janet Museveni, wife of President Yoweri Museveni, opened the exhibit and the week. She asked many questions, and was fascinated by the chimpanzees' use of medicinal leaves (observed in Uganda as well as Tanzania, and involving bitter foliage not normally eaten). She arranged for me to meet her husband. "I understand your love of chimpanzees," he told me. "For I feel the same about my cows. When you next come, I must take you to see my cows."

In Burundi the minister for the environment, Louis Nduwimana, opened the exhibit. He lingered at "The Last Tree." This is a painting based on a dream of my mother's, in which she saw the destruction of a forest and the disappearance of all but one family of chimpanzees. Finally he turned to me. "Now," he said, "I understand your passion."

In the Republic of Congo, a young woman told me that her family had a chimpanzee when she was growing up. They loved him, and dressed him up, and then he got pneumonia and died. "I didn't understand until now how cruel we were to keep him," she said. Her eyes were full of tears and, as she turned away, I saw them spill down her cheeks.

In Sierra Leone the head of state himself, Capt. Valentine Strasser, opened the exhibit and asked many questions. Later I had several meetings with a senior official, Lt. Col. Abdul Karim Sesay. When he was a young man, he said, he had been given a juvenile chimpanzee whose mother had been shot in the nearby forest. Three times the chimpanzee escaped from his cage, and twice he was caught and put back. "The third time I told them to leave him. Why should I force him to stay with me when he wanted to go back to his own family?"

All over the world people react with amazement when they see slides and film of chimpanzees' family life, their use of tools, and their nonverbal communication patterns—kissing, embracing, holding hands. Stories of the Gombe community—the death, in a high nest, of old Melissa as I sat grieving below; the overthrow of Goblin after a nine-year reign; the adoption of sickly Mel by the adolescent male Spindle—move people strangely. "I never would have believed I could become so emotionally involved with a family of wild animals," a woman told me in Germany. "Are you going to write another book?" asked an elderly man in Edinburgh, Scotland. "I want to know how Flo's family is getting on."

The thing is, chimpanzees do not behave as people feel animals should behave. They make us realize that we are not quite so different from the rest of the animal kingdom as we used to suppose. The great apes do, indeed, occupy some hazy area where two worlds meet, the world of human beings on one side, nonhuman beings on the other. They make us humble. They teach a new respect for all the amazing nonhuman beings with whom we share the planet.

For me, 32 years ago, it all began with David Greybeard. Calm, less fearful than the others, it was David who first allowed me close, who demonstrated the use of stems and twigs to fish for termites, who showed that chimpanzees, though eating mostly fruits and vegetation, love insects too. It was he who first showed me that chimpanzees eat meat. And it was David who, on a far-off day three decades ago, took a palm nut from my hand. While unimpressed by the fruit, which he dropped, nevertheless he showed me that he understood, appreciated my gesture in offering it—he took my outstretched hand in his and held it gently. If I close my eyes, I can still feel the firm pressure of his smooth, warm fingers. We had communicated in a language far more ancient than words, a language that bridges two worlds.

A History of Research

Hand in hand through the forests of Borneo, primatologist Biruté Galdikas takes her orangutan charge for a walk on the wild side. Supported in part by grants from the National Geographic Society, scores of scientists have brought new light to bear on the daily lives of the great apes—confirming the words of the Roman poet Ennius: "Ugly brute as he is, how like us is the ape!"

By Mary G. Smith

23

Istared at his ferociously scarred face through the bars of the cage, a sight I recall without effort. I noticed his eyes especially—they seemed so human and sad. I felt so sorry for him. Finally my mother dragged me and my little sister away to see the rest of The Greatest Show on Earth, but I never forgot Gargantua. It was 1940. I was six years old, and from that day on I was hooked on apes.

Gargantua was a lowland gorilla, brought from Africa to the U.S. by ship as a youngster in 1931. A disgruntled sailor squirted nitric acid into his face, permanently damaging his facial muscles and the skin around his eyes. His owner, a charmingly eccentric Englishwoman married to a Brooklyn doctor, called him Buddy and cherished him until he grew too big and unpredictable for safety—the usual story. Finally, reluctantly, she sold him to Ringling Brothers and Barnum & Bailey Circus. Renamed and ballyhooed, he awed and fascinated circus-goers until his death from pneumonia in 1949.

Fast forward to 1961. Two men walk into my office at the National Geographic Society in Washington, D.C. "How old are you, my dear?" "Twenty-seven," I say, startled. "Good, good," he says, beaming and rubbing his hands. "She'll like that. She's the same age."

The "he" was Dr. Louis S.B. Leakey, the late, great paleoanthropologist from Kenya. The "she," I found out later, was Jane Goodall, a young Englishwoman in her second year of studying wild chimpanzees on the shores of East Africa's Lake Tanganyika. Leakey and Dr. Melville Bell Grosvenor, then the Society's President and Editor, had recently discovered each other, the beginning of a warm and fruitful friendship. Funded by the Society for research into the origins of humankind, Leakey had also talked its Committee for Research and Exploration into contributing a little money to keep Goodall in the field. Both Grosvenor and Leakey were world-class enthusiasts about anything that interested them, and most things did.

Three decades ago those two men in large measure set the course of my life. I was a NATIONAL GEOGRAPHIC picture editor hired by Grosvenor in 1956. Leakey wanted someone—preferably another female—to meet Goodall and plan a photographic coverage of her work. Grosvenor picked up the idea with enthusiasm. As Editor he believed in strengthening his staff's field experience; picture editors were frequently sent on trips.

Within a few months I was off to Nairobi, frantically reading Robert Ardrey's *African Genesis* and Isak Dinesen's *Out of Africa*. Virtually everything I knew about Africa when I stepped off the plane into the brilliant Kenya sunshine came from those two books. Much of the world knew little about Africa and even less about great apes in the wild. Field research into the lives of our closest living relatives was in its infancy. In 1959 George Schaller had begun his pioneering study, 20 months with mountain gorillas in Uganda and what's now Zaire; he reported his work in two of his beautifully written books, *The Mountain Gorilla* and *The Year of the Gorilla*. Thanks to Schaller the huge, gentle ape began to shed its reputation as a terrifying despoiler of the coun-

Impresario of ape research and famous for his own discoveries of African fossils in man's lineage, Louis S.B. Leakey encouraged long-term field studies of the great apes—studies that would reshape the gargantuan myths often associated with these animals.

tryside, a marauding beast bent on the destruction of everything in its path, especially humans. How did the gorilla come to rate such a ferocious billing in the first place?

Even in antiquity, reports filtered in to Mediterranean cities about large apes sighted and even captured. In 470 B.C. an expedition of colonists left Carthage in a fleet of fifty-oared galleys. On the West African coast they encoun-

tered hairy, stone-throwing creatures they called *gorillai,* perhaps from a local word. Two thousand years passed before European adventurers and explorers added more information. In 1774 a British sea captain reported a "wonderful and frightful production of nature....7 to 9 feet high.... thick in proportion and amazingly strong...."

Other accounts, some believable and some not, surfaced here and there until 1846, when an American missionary named Thomas S. Savage

A moment of trust cements a bond of friendship as Flo lets her baby, Flint, greet Jane Goodall early in 1965. Here the infant chimpanzee cautiously examines the proffered hand; Flo would understand its back-of-the-wrist pose as a gesture of submission. From three decades among the chimpanzees of Tanzania's Gombe National Park, Goodall's observations revolutionized thinking about these animals. She observed their ability to make and use tools— and recorded all-too-human acts of violence and predation.

discovered a gorilla skull in the home of a Reverend Mr. Wilson, beside the Gabon River. It was sent to anatomists Jeffries Wyman and Richard Owen with hair-raising descriptions that characterized gorillas for a century. Savage wrote: "They are exceedingly ferocious, and always offensive in their habits, never running from man as does the Chimpanzee.... It is said that when the male is first seen he gives a terrific yell that resounds far and wide through the forest....he then approaches the enemy in great fury, pouring out his cries in quick succession." If the man falters or his gun misfires, "the encounter soon proves fatal to the hunter."

In 1856 an American journalist, Paul Du Chaillu, traveled to West Africa and became the first person known to shoot a gorilla. It is said that his publisher made him exaggerate the dramas of his book *Explorations and Adventures in Equatorial Africa*. His first gorilla "reminded me of nothing but some hellish dream creature—a being of that hideous order, half-man half-beast...in some representations of the infernal regions"—an ogre killed at six yards' range.

As I mull over these bloodthirsty tales, Dian Fossey strides into my memory—tall, eloquent, dedicated. Also quirky, obsessed, self-destructive. Dian Fossey, the savior of the mountain gorilla. We knew each other for almost twenty years. I first heard her name in a telephone conversation with Louis Leakey in 1966. He called me in great excitement to say that "the perfect person is going to Africa to work with mountain gorillas!" He had asked the Society for funds for her. Would I talk to her about photography? Didn't I think her future work would make a wonderful article? The usual cascade of Leakey enthusiasm poured from the phone. "So, do we have another Jane Goodall on our hands?" I asked. "Yes, yes, but this one's dark. And much taller. She's going to be wonderful." And so she was, in her own weird way.

The last time we ever saw each other, Fossey leaned against the door of my office, looking good, I thought, but almost serenely weary. It was 1983. She had spent a three-year "exile" in the States, a banishment forced on her by the U.S. and Rwandan governments, her supporters (the Society among them), and the scientific community. Her volcanic temper and unpredictable, bullying behavior with humans had raised fears for her safety. She had accepted a visiting professorship at Cornell, and completed her book, *Gorillas in the Mist*. Now she was on her way back to Rwanda and her beloved gorillas.

"Aren't you ever going to visit me at Karisoke?" she asked. "Sure," I said. "One of these days when I have time. When are you coming back to the States?" "I'm not. Not ever again. If you want to see me, you'll have to come to Rwanda."

At Christmas 1985 she was murdered, slashed to death in her cabin at Karisoke, her research center in the Virunga Mountains. Her killer remains

unknown. In August 1986 the Christmas card I'd sent her landed back on my desk. On the envelope, in French: "Deceased. Return to sender."

When Fossey went to the Virungas in 1967, it was thought that fewer than 250 mountain gorillas survived in the wild; none has lived for any time in captivity. The Virunga population grew in the 1980s. By 1993, an expert estimate, including 300 in Uganda's Impenetrable Forest, put the total around

WOODCUT FROM PAUL DU CHAILLU, *EXPLORATIONS AND ADVENTURES IN EQUATORIAL AFRICA*, LONDON: JOHN MURRAY, 1861

600. Fossey's true contribution was to save them. I discussed this with George Schaller not long ago, and he sent me a review he had written about one of the numerous books on Fossey that appeared after her death.

"She made several important new observations on gorilla behavior," he noted, "including the discovery of infanticide and the transfer of females out of old-established groups to new ones." But, he said, "after her favorite gorilla, Digit, was killed by poachers in 1977, she abandoned all pretense of scientific effort....With singular devotion, Dian made the correct choice for herself: gorilla protection had to take precedence over research....she helped this magnificent ape endure during a critical decade of its history."

Every gorilla now seen in the flesh in captivity is the lowland variety, a bit smaller than its mountain cousins and with shorter hair, less girth of chest, and less pronounced skull crest. For centuries it has been hunted by humans for food, and is extremely wary throughout its range in Central and West Africa. Scientists in the Central African Republic and Gabon have worked on them for years, and J. Michael Fay of NYZS/The Wildlife Conservation Society (until early 1993 the New York Zoological Society) has recently moved to the Nouabalé-Ndoki region of Congo.

I asked George Schaller, whose own ties with NYZS began in 1966, about Fay's research. "Almost no sustained eyewitness observations have been possible either in the CAR or in Gabon. Lowland gorillas are much more arboreal than mountain gorillas, so the research consists mostly of feces collecting and analysis. We've gotten some solid census data but little else except frustration. In the Ndoki region, Mike Fay'll be in contact with gorillas and chimpanzees that have rarely or maybe never seen humans. They aren't so leery of us, so maybe long-term observations can begin."

So far the prize for long-term field studies goes hands down to Jane Goodall for her work over three decades. Although her work has been supported by hundreds of private individuals and organizations other than the Society, there is never, ever, enough money. Circling the globe, giving lectures, meeting donors, she pleads, wheedles, cajoles—and inspires. Respectfully underawed by officials, she has gone to heads of state and captains of industry, and won their cooperation by dint of warmth, friendliness, and cast-iron will. Her time at Gombe amounts to two or three days a month, and well-trained assistants carry on day-to-day observations.

But how did that tradition really begin? With a man in a cage in Gabon in 1893, believe it or not. R. L. Garner, an American zoologist, assembled his cage in the jungle—to protect himself from the fierce great apes—and spent 112 days and nights behind bars. Most animals were far too wary to approach this peculiar habitat, so he saw little. Much of his information came from natives, supplemented by his own theories and imagination. Nevertheless, his book *Apes and Monkeys, Their Life and Language*, published in 1900, is an entertaining treasure, the hard-won musings of a scientist wannabe. He was the first person to conduct a serious field study of great apes by observational methods, not by killing his subjects first and studying them later.

In 1930 the renowned psychobiologist Robert Yerkes, who was studying captive chimps in his laboratory in Florida, sent Henry Nissen to West Africa for a dry-season field study—49 days of actual observations. By 1943 Yerkes could write, in a summary of knowledge about chimpanzee intelligence, that elements of learning in these apes would probably soon be "identified as antecedents of human symbolic processes....discoveries of moment seem imminent."

"And here, just as he began another of his roars, beating his breast in rage, we fired, and killed him." So wrote Paul Belloni Du Chaillu, an American of French extraction and the first white known to bag a gorilla. His book Explorations and Adventures in Equatorial Africa, *published in 1861, proved an instant success. But its overblown prose and exaggerated descriptions presented the gorilla as a demonic monster, "some hellish dream creature ...half-man half-beast"—an impression that would persist well into the 20th century.*

BOB CAMPBELL

Dian 1970.

*The gorillas "are the reward,"
wrote Dian Fossey in 1976, "and
one should never ask for more than
their trust...." Trust her they did.
From 1967 until her untimely death
in 1985, Fossey devoted herself to
the mountain gorillas of Rwanda,
studying their family relationships
and dispelling notions about their
ferocity. In scenes annotated
by photographer Bob Campbell,
she scans the forest for gorillas
and examines their bones
in her hut high on Mount Visoke.*

At about the time Jane Goodall set off for Gombe, in 1960, Japanese scientists began a survey in Tanzania. This would turn into Toshisada Nishida's long-term research project in the Mahale Mountains. Early in 1960, Adriaan Kortlandt from the Netherlands began observations from a pawpaw plantation in the Belgian Congo. Two years later, in a landmark article, Kortlandt wrote: "Unlike zoo chimpanzees, which generally look increasingly dull and vacant with the years, the older wild chimpanzees seemed to me more lively, more interested in everything and more human." They gave him the feeling of "looking at some strange kind of human beings dressed in furs.... Once I saw a chimpanzee gaze at a particularly beautiful sunset for a full 15 minutes...."

Goodall's first accomplishment was simply to get near the chimpanzees without scaring them away. Then astonishing news began trickling in from her camp. She saw chimpanzees altering twigs and grass stems—in other words, making tools—to poke into termite mounds. After a minute or two, the chimps would withdraw them and lick off the clinging insects.

A few years later, one of her students, Geza Teleki, now a well-known anthropologist and chairman of the Committee for Conservation and Care of Chimpanzees, tried to find out if humans could do as well. He spent much time observing a chimp named Leakey and trying to master his expert technique. Teleki wrote a scientific report: "Despite months of observing and aping adult chimpanzees as they selected probes with enviable ease, speed and accuracy, I was unable to achieve their level of competence. Similar ineptness can only be observed in chimpanzees below the age of about 4-5 years."

In 1962, outside a restaurant in downtown Nairobi, I laid eyes on Jane Goodall for the first time. I was in Kenya to meet with Louis Leakey and try to figure out how to illustrate his research into early human fossils. I was also under orders from Melville Grosvenor to find a way to do an article about "the blond British girl studying the apes." As she and I shook hands it was sprinkling rain, and she was wearing a blue-and-white cotton dress. She wasn't much interested in talking to me, I recall, and I wasn't particularly fascinated by her, either. I sent a letter to a close friend back at the office, saying that Jane seemed nice but sort of frail and that she probably wouldn't last. So much for my ability to read the future.

I also arranged for a young Dutch photographer, Hugo van Lawick, to make the pictures for Louis Leakey's forthcoming articles—and also to cover Goodall's work. He produced, I would say, the most remarkable collection of photographs of wild animal behavior ever seen until then. He was also a talented cinematographer, and lecture and television films as well as GEOGRAPHIC articles started Goodall on her road to fame.

Goodall and van Lawick were married in 1964 but divorced ten years later. Her second husband, Derek Bryceson, died within five years, felled by cancer, and she immersed herself even further in her work. What does she think is the biggest achievement of her long career?

"For more than three decades, I've been able to watch infants grow up," she says. "The ones with supportive, affectionate mothers turned out to be confident, high ranking, and assertive. Mothers who were rejecting and nervous tended to produce offspring who were jumpy and who had diffi-

BOB CAMPBELL

Dian in 1st cabin – 1969.
Studying gorilla bones.

culty entering into calm, relaxed relationships. Humans can hide the effects of the small traumas of early life, but chimpanzees act the way they feel."

Louis Leakey was sure that studies of living apes in the wild would illuminate questions about early human—or pre-human—behavior. An important multiyear field study in West Africa bears this out. Since 1979 Swiss biologist Christophe Boesch and his wife, Hedwige Boesch-Achermann, have observed the chimpanzees of the Taï Forest in Côte d'Ivoire. These chimps use a hammer-and-anvil technique to free nutmeats from their hard outer shells. Mothers patiently teach their offspring how to pound a nut with enough finesse to free a perfect, unsquashed kernel—a skill that takes ten years to perfect, and a striking example of tool use passed from one generation to another. Boesch has also reported frequent cooperative hunting, and food sharing after a kill.

Counterparts of these practices are common among humans. In fact, Frans de Waal, of the Yerkes Regional Primate Research Center in Atlanta, says, "Humans have more developed culture and language, but we are seeing that our social system is an ape social system."

Writing in the March 1992 issue of NATIONAL GEOGRAPHIC, Eugene Linden asks, "What had happened by roughly five million years ago to cause Africa's tree-living apes to descend to the ground, then develop into divergent lines—modern apes and humans?" Linden, author of three books on ape language research, points out that climate probably affected two key events: descent to the ground and our ancestors' increasing reliance on their brains for survival. "We tend to romanticize human evolution in heroic terms," says Linden, "but a prosaic event such as a change in diet may have driven the apes to develop large brains.... If we could understand why apes have bigger brains than other primates, we might discover why humans have larger brains than apes."

An ape that's helping shed light on these questions is the bonobo, or pygmy chimpanzee. Years ago I was working with Randall Susman, an anatomist at the State University of New York at Stony Brook. We were looking at color transparencies that he'd taken during a Society-sponsored project in Zaire. Suddenly a strange-looking ape popped up on my screen.

"What in the world's that?" I asked him. "Is it a chimp? It looks odd." "Yes," Susman replied. "It's a chimp, but it's one that almost no scientist ever sees. It's called a bonobo—a pygmy chimpanzee. We don't know much about them yet."

The creature that stared back at me looked like no chimpanzee I'd ever seen before. More graceful in appearance and more slender than the so-called common chimpanzee, it also had smaller ears and a Beatle-like head of hair. Now found only in the rain forest of central Zaire, bonobos weren't identified as a separate species (*Pan paniscus*) until 1933, and it's estimated that fewer than 20,000 still exist.

Unlike the other great apes, bonobos sometimes walk upright, they live in groups of both males and females, and they're usually peaceable and friendly. Females are on an equal footing with males, and sexual behavior is part of many social interactions—bonobos seem to have adopted a make-love-not-war philosophy, saving themselves a lot of trouble.

For years the Japanese have been in the forefront of bonobo field studies. Randall Susman has worked at Lomako in Zaire, the only "Western site." In 1972 Toshisada Nishida made a bonobo survey in Zaire, and a year later Takayoshi Kano—on a bicycle, no less—searched throughout north-western Zaire for a suitable permanent study area. Kano finally decided on a site near the village of Wamba. He has worked there ever since with the help of the local Bagandu people, learning their language and hiring them as assistants and trackers. Thanks to a Bagandu legend about human ancestors who once lived as brothers with the bonobos, the apes have been safe from hunting and aren't as wary as bonobos elsewhere. Since they gather in large groups to feed, they're good targets for trappers and hunters. But as always with apes, the biggest threat is human incursion—loss of habitat through population growth, land clearing for farming, and logging. As long as field research continues at Lomako and Wamba, the bonobos are relatively safe. If the work is cut off by lack of funding or because of civil unrest, they'll probably disappear very quickly.

Technical journals excepted, little has been published in the West on the work of the Japanese primatologists, but for decades they've studied various primates—notably macaques, bonobos, and chimpanzees—and contributed greatly to scientific knowledge. Unlike Western specialists, who by tradition take a biological standpoint, the Japanese approach their subjects as cultural groups. They have not hesitated to attribute a reasoning mind to their research animals, and this has allowed them to form a realistic picture of primate societies as groups of individuals with differing and not always predictable behavior. Indeed, important research centers in Japan hold annual memorial services for their subjects. At these ceremonies, laboratory primatologists pray for the souls of animals that have died during the year.

Until Jane Goodall came on the scene, the idea that nonhuman primates might have minds and even souls worth worrying about had no parallel in Western science. It would have been labeled sentimental, overly emotional. Alison Jolly of Princeton University wrote a useful summary in her review of Goodall's 1986 monograph, *The Chimpanzees of Gombe:* "Fashions and preoccupations in primatology have changed over the decades. It may be interesting to recall how some themes of Goodall's work struck the rest of us and how our attitudes have grown. The theme of the individual's importance started early. Goodall named her chimps as soon as she could recognize them. At that point only the Japanese admitted to being sensitive to indi-

Pioneers of primate research, Russia's Nadie Kobts and America's Robert M. Yerkes established psychological similarity to humans in the 1920s. Mrs. Kobts, here in Moscow with her chimp Ioni, revealed that these animals see the same shapes and colors humans do. Yerkes studied chimpanzee intelligence, noting that Prince Chim (right), a pygmy chimp, or bonobo, had exceptional intellectual powers. "Seldom daunted," wrote Yerkes, "[Chim] treated the mysteries of life as philosophically as any man."

HENRY BURROUGHS,
PHOTOGRAPHER/COURTESY LIFE MAGAZINE

*Stand-in for an astronaut, Ham,
a three-year-old chimpanzee,
awaits release from his seat after
a 1961 ride into space aboard
a Mercury capsule. The 16-minute
flight carried him 156 miles high
and 414 miles downrange at speeds
as great as 5,800 miles an hour.
Science and entertainment
have made apes an integral
part of modern life—whether
in laboratories or through scenes
such as Fay Wray in the clutches
of Hollywood's King Kong.*

vidual primate temperament, and they admitted it to each other in Japanese....
In the early 1960s, traditionally trained ethologists wondered how seriously
to take Goodall, even when she promised to come in from the wild and write
a doctorate at Cambridge. Now Goodall's approach has been fully justified."

A few primatologists are less kind about my longtime friend Biruté Galdikas, the third and youngest of the "ape ladies." Louis Leakey
sent her to the swamp-ridden forests of Borneo to find and study Asia's
great red ape, the orangutan, and scientists and laypersons alike respect
her for her bravery and dedication. Conservation has claimed a share of
her energy from the first. Her years of sustained work have let her discover
such significant facts as the typical interval between orangutan births—
an astonishing eight years—and the foods eaten by the apes in her 14-
square-mile study area, no less than 400 different kinds, including insects.
In the last decade a few of her peers, anxious for a full scientific account
of her study, have criticized her for not publishing more (much as Goodall
was criticized before her monograph appeared). Now Indonesian officials—
with an eye toward road building and future logging in her study area,
Tanjung Puting National Park—strew her path with red tape when she
tries to renew her research permit. Without a permit, she can't return to her
base, Camp Leakey, and her twenty-year effort to study and protect the
orangutans would come to an abrupt end. Then, by a cruel irony, her subjects could well be on their way to oblivion. For years she has dug in
her heels and concentrated on looking after the apes (for part of the time
with earnest ecotourism assistance).

What's the solution to a vicious circle like this? To analyze and evaluate their observations, most scientists conducting long-term studies must leave
the field and concentrate on writing for publication—but as soon as they
leave, their subjects come under increased threat. The great apes are found
in countries where annual incomes are stated in hundreds of dollars or less,
humans are crowded and hungry, and life is short and tough. Governments—
however enlightened about their national nonhuman living treasures—put
the needs of humans first.

What happens when the scientist leaves? Jane Goodall puts her trust
in a well-trained staff. Dian Fossey, forced to revisit the U.S., left her research
center in the hands of fellow scientists and Rwandan assistants who looked
after things reasonably well—a fact that she only grudgingly acknowledged.
Others rely on local folk, and hope.

But without scientific publication, major sources of funding dry up.
Researchers spend more and more time seeking money, spirit withering, outlook gradually souring. Of course, this ordeal has different effects on different people. Over the years Goodall has become disarmingly serene. Fossey
very nearly went mad. And Biruté Galdikas, when I saw her recently, was
quietly frantic.

Seeking money for her project and help from her friends to persuade the Indonesian government to renew her permit, Galdikas came to my office with 16-year-old Binti, the son born during her marriage to Rod Brindamour. Like Galdikas a Canadian citizen, Brindamour was the project's photographer, surveyor, fellow observer, and general manager, and his pictures let me follow Binti's childhood. In 1980, on a GEOGRAPHIC cover, Binti shared a dishpan bath with Princess, a tiny infant orangutan—a hugely popular picture. Even

today we're asked, "Whatever happened to the baby and the red monkey?" (The answer: "It's an ape, not a monkey, and the baby's a student in Canada.")

When Galdikas and Brindamour went to Indonesia, two of the islands, Borneo and Sumatra, were the only homes left for an estimated 20,000 orangutans. These were the last remnants of a population that once numbered in the millions and extended from central China in the north to Java in the south and as far west as India. In the 13th century Marco Polo's accounts of his travels to the Orient spoke of apes, probably orangutans but possibly gibbons, the lesser apes of Asia. The earliest undisputed description came in 1658 from Jacob de Bondt, or Bontius, a Dutch physician working in Java. He was first to apply correctly the word "orangutan"—Malay for "person of the forest"—but described the creature as a "wonderful monster with a human face." He also noted that the Javanese claimed these apes could speak but didn't, "lest they should be compelled to labor."

Hearsay hasn't made modesty a virtue of male apes, and a later scholar says, "There were actually rumors in Borneo that male orangs carry off the native women for sexual purposes."

Even the great naturalist Linnaeus had to rely on unreliable reports. (Known as Carl von Linné to his fellow Swedes, he devised the standard system of Latin scientific names that botanists and zoologists use today.) In his *System of Nature* in 1758 he put the orangutan in the same genus as humans, calling it *Homo nocturnus* or *Homo sylvestris orang-outang*. His account says: "Body white, walks erect, less than half our size, hair white, frizzled . . . . Lifespan twenty-five years. By day hides; by night it sees, goes out, forages. Speaks in a hiss. Thinks, believes that the earth was made for it, and that sometime it will be master again, if we may believe the travellers."

In the 19th century many travelers visited Borneo to find out more about the apes—and collect specimens. Alfred Russel Wallace, who formulated a theory of evolution very like Darwin's, shot several of the great red apes. In 1878 William Temple Hornaday, chief taxidermist of the United States National Museum, killed 43—seven of them in one day. Later he wrote: "I would not have exchanged the pleasures of that day, when we had those seven orangs to dissect, for a box at the opera the whole season through . . . . When we finished, there was a mountain of orang flesh, a long row of ghastly, grinning skeletons, and big, red-haired skins enough to have carpeted a good size room."

Such specimens, at least, could serve a developing knowledge of primate anatomy. Most just became trophies for sportsmen happily snuffing apes—and other game. Tom Harrisson, of the Sarawak Museum, wrote that by 1900 orangs "were no longer readily available for the Anglican priest with a shotgun and Monday off or the naval officer with short shore leave and the longing to kill on land." His wife, Barbara, took orphan orangutans into her home, studied their behavior, and tried to reintroduce them into the wild.

Biruté Galdikas inherited a similar responsibility in 1971, when the

Indonesian government was beginning to enforce a law against killing or even owning orangutans. She agreed to accept animals confiscated by the authorities, rehabilitate them, and set them free. Eventually most returned successfully to the forest, but not before they made a daily shambles of the camp, playing, looking for food. In the October 1975 GEOGRAPHIC, Brindamour's pictures and Galdikas' text recorded affectionate struggles with the young rehabilitants: "We often woke to find not one, but four orangutans in bed with us.... Our thatch was leaky from the holes caused by orangutans walking along the roof.... During heavy rains, everything in the hut got soaked."

After eight years the couple had, among other accomplishments, logged more than 12,000 hours of observation of wild orangutans. Brindamour, however, decided to return to Canada and his interrupted career in computer science. Divorce and remarriage followed. Galdikas married Pak Bohap bin Jalan, a Dayak tribesman, and I expect that they will continue her life's work: saving the orangutans of Tanjung Puting National Park—and, I hope, telling more of their story.

━━━━━━━━

If apes could talk to humans, what would they tell us about themselves? What would they ask about us? A few years ago I finally met—in person, one could say—a longtime collaborator. I'd worked as an editor with her and her colleagues since the mid-1970s, but this was the first time we'd seen each other. As I sat down beside her in the kitchen of her comfortable trailer home, she wanted to know right off the bat what was in my purse. Did I have a lipstick? A mirror? Could she see the mirror? Could she look at my teeth? Odd questions for a first meeting, but then Koko's a fairly unusual questioner. She's a lowland gorilla, winsome for all her 270 pounds, and very smart. We stared at each other for a second. Then she reached for me with her huge hands, fumbled with the buttons and zippers on my clothes, and gently stroked my hair. Carefully she put a massive arm around my shoulders. I flinched and tried to back away, but it was like pushing against a brick wall. I was trapped, and we both knew it. I decided to relax and be grateful that I was dealing one-on-one with curious, fidgety, good-natured Koko and not with Gargantua.

Koko has an ASL—American Sign Language for the deaf—vocabulary of hundreds of words and can put them together in rudimentary sentences. Working with psychologist Francine "Penny" Patterson and Patterson's long-time partner, molecular biologist Ronald Cohn, Koko has been for about twenty years at the center of a storm: of curiosity and something like a movie star's adulation from the public, skepticism from many scientists. None of this bothers Koko, but Patterson has been accused of overinterpreting Koko's rapid-fire gestures and of publishing too little for the scientific community. Patterson contends that her peers don't make an effort to understand what she's

VOL. 154, NO. 4   OCTOBER 1978

NATIONAL GEOGRAPHIC

OFFICIAL JOURNAL OF THE NATIONAL GEOGRAPHIC SOCIETY WASHINGTON, D.C.

accomplishing. "At one time or another I've invited them all to visit us to see what we're doing," she told me last spring, "but they never come." Critics claim that a request for a visit with Koko would meet evasion and probably permanent delay.

To avoid disruptions in training, Patterson and her assistants spend virtually all their waking hours with Koko, with Michael, Koko's consort, who's also proficient in sign language, and with a younger male gorilla, Ndume. In the GEOGRAPHIC for October 1978, she told of Koko's progress. She stressed the importance of displacement—the ability to refer to events removed in time and place, in human language—and quoted "a revealing conversation" about an incident three days earlier, when Koko had bitten her: "Koko: 'Wrong bite.' Me: 'Why bite?' Koko: 'Because mad.' Me: 'Why mad?' Koko: 'Don't know.'" This, said Patterson, conveyed more than an animal's "dim, unsorted recollections of pain and pleasure."

From the start of the project, Ron Cohn had won Koko's trust. A scientist, he's also a talented photographer with an excellent editorial sense, and he was able to capture meaningful sign-language illustrations with still film. Lack of such pictures, after a couple of years of trying, had defeated our effort to produce an article on the first researchers who attempted to teach ASL to an ape, in the 1960s.

These were psychologists R. Allen Gardner and Beatrix T. Gardner, of the University of Nevada at Reno, who taught scores of signs to a female chimp named Washoe. Their pioneering work inspired a tradition, and ape language projects have survived in spite of ferocious political in-fighting, and a general drying up of funding in the late eighties.

In 1992 the Society did approve a grant to Sue Savage-Rumbaugh, who works with chimpanzees and bonobos at the Language Research Center near Atlanta. Her background in the developmental psychology of human children, her careful experiments, and her willingness to share her results have helped to restore faith in such projects. She and her colleagues have taught a complex collection of lexigrams, geometric symbols with set meanings, to a precocious bonobo named Kanzi. He combines symbols for actions and objects: "hit" and "ball" or "tickle" and "Kanzi." Apparently he also understands that a change in the order of words can signify a change in meaning.

Robert Yerkes wrote, in 1925: "I am inclined to conclude... that the great apes have plenty to talk about, but no gift for the use of sounds to represent individual... feelings or ideas. Perhaps they can be taught to use their fingers...."

How right he was, this early giant of primatology. But will we ever be able to converse easily with our nonhuman kin, or they with us? And if we never can, will the experiments have been worthless? Do they have lasting value? Of course they do. Consider what the language-proficient apes represent, with their specialized learning and their peculiar hybrid existences. These chimpanzees and gorillas and bonobos are a living interface between

ROD BRINDAMOUR

Cover stories: Biruté Galdikas, Louis Leakey's third "ape lady," accompanies a brace of orangutans through the wilds of Borneo. Since 1971 she has rehabilitated many of these wild-born animals, most of them pets now being returned to the forest. Koko, the world's first "talking" gorilla, snaps a self-portrait in a mirror. Taught by researcher Francine Patterson to communicate in sign language, she responded "Love camera" when shown a copy of this picture.

the human and nonhuman: a little less ape and a tiny bit more human than their wild counterparts. At the very least they're ambassadors from their world to ours, and they should be treated—if not as equals—with consideration and respect. Our recognition of their right to exist with dignity makes better human beings of us all.

Nowhere is this right seen more clearly than in the work of a group

of genuine heroes, extraordinary and basically unsung: the field photographers who've worked under arduous, frequently unspeakable conditions for years with the famous researchers. The observations the scientists made and the words they used taught us new and amazing facts. But the pictures seized our imaginations and our hearts and made believers of us in an instant. Without the courage and talent of people like Hugo van Lawick, Alan Root, Bob Campbell, and Rod Brindamour, with a dedicated next generation represented by Frans Lanting and "Nick" Nichols, this book couldn't exist—and neither would many of the field studies.

So what's the common denominator among Leakey's "ape ladies," Goodall, Fossey, and Galdikas? Why did Leakey seem to select only females for field studies? The explanations boil down to the theory that he believed women were better than men at patience, persistence, and perception. Another element is important—he just plain liked women. I knew Leakey well for years. Our friendship survived many things, including his offering me— and my rejecting resoundingly—the chance to give up life and career in the U.S. to study, of all things, aardvarks! He trusted women, felt easy around them, knew how to inspire them. Both Goodall and Galdikas will insist today that he influenced their lives for their good.

All three had read everything available about their particular species of ape, but only Galdikas had any formal training in primatology. Goodall and Fossey began with little to fear from academia—no peers or advisers to be offended by a new discovery or theory. They could publish what they chose. In time all three earned doctoral degrees at prestigious universities, and now Goodall and Galdikas lecture or teach at least part of the time. They advise young women and men who yearn to follow them to begin with a solid education—biology and physical anthropology are must courses—and plan for an advanced degree. It also helps to accept the fact that most primate research is not romantic or glamorous.

Well, Dian, almost a decade ago—the last time I saw you alive—you asked me to come to Africa to visit you and the mountain gorillas. Now, late in 1992, I set these words down by candlelight at your old camp at Karisoke, high in Rwanda's Virunga Mountains. At last I climbed the slippery, almost impossibly difficult trail, labored through the green and dripping forest to meet your gorillas, and once again said hello and good-bye to you. In a clearing near a massive *Hagenia* tree I read the bronze plaque on the cairn of volcanic stones that marks your grave:

"Nyiramachabelli" / Dian Fossey / 1932-1985 / No one loved gorillas more / Rest in peace, dear friend / eternally protected / in this sacred ground / for you are home / where you belong /

In marked graves nearby lay sixteen gorillas—Nunkie, Marchessa, Uncle Bert, your beloved Digit, and a dozen others—victims of disease, age, and like you, murder. The sun broke through the clouds for a few seconds, and the raindrops sparkled on the gray moss draping the trees. A tiny antelope—a black-fronted duiker—peered through the tall grass, then bounded away. A faint breeze rose and quickly died. I heard your odd, whispery voice again. I saw you everywhere I looked. You lost your life, but you saved the mountain gorilla. Now your heart and your hopes and your legacy belong to the world forever. The great apes are so much like us, very nearly our sisters and brothers. You showed us that we can and must be their keepers.

Nyiramachabelli: To Dian Fossey, this African nickname meant "the old woman who lives in the forest without a man." Here it embellishes her tombstone where she rests in the company of her beloved gorillas—whose portraits cover a wall of her nearby cabin.

Gibbon
Hylobates species

U nlike other primates, apes and humans lack tails. The lesser apes, the gibbons of Southeast Asia, are the smallest: 8 to 25 pounds. They swing through the trees, walk upright along limbs.

Today only an estimated 20,000 orangutans remain in Sumatra and Borneo. Adult males have bulging cheek pads and can weigh 200 pounds.

Africa's bonobos, or pygmy chimpanzees, were formally

IN ALL APE PAIRS ILLUSTRATED, THE FEMALE IS ON THE LEFT.

Orangutan
Pongo pygmæus

Bonobo
Pan paniscus

PAINTING BY ROBERT E. HYNES

recognized as a species in 1933. Slighter and a bit smaller than common chimps, they walk upright more often and live in sexually active groups as large as 40. Males weigh about 100 pounds, females 70. Their range is a single rain forest in Zaire, while common chimpanzees exploit more varied habitats. Males, larger than females, can reach 110 pounds. More aggressive than bonobos, chimpanzees live in smaller, male-dominated groups. Males compete for status and one will kill another—or his offspring.

Dignified vegetarians, gorillas are the largest primates. A silverback male might weigh 400 pounds, his mate half as much. Both mountain and lowland gorillas are in danger of extinction in the wild.

Adaptable, omnivorous, humans have become bipedal walkers with longer legs, shorter arms, and a lower spine curved to place the center of gravity over the pelvis. The brain is larger and more complex. Anatomy affected by the upright carriage of the head makes it possible

to produce the sounds crucial and peculiar to human speech. Like those of other primates, human young must learn social and survival skills through a lengthy childhood and adolescence.

Chimpanzee
Pan troglodytes

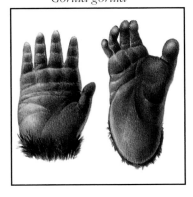

Gorilla
Gorilla gorilla

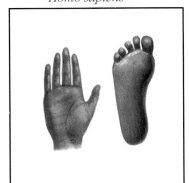

Human
Homo sapiens

The Kingdom of the Wild

"A primeval forest of marvellous beauty," a visitor in the 1920s described the home of the mountain gorilla, in Central Africa. A vigorous adult male forages in a giant senecio. In recent decades long-term studies of great apes—gorillas, bonobos, chimpanzees, orangutans—free in their wild environments around the world have revealed many surprising aspects of their lives.

By Michael Nichols

45

S he's down there," said the bush pilot, "the crazy American woman who lives alone with the gorillas." She, of course, was Dian Fossey, and down there was her research area. I knew about her work, and other ape projects, because I had grown up reading NATIONAL GEOGRAPHIC, and in 1980 I made my first effort to photograph gorillas.

It was a very careful experience. The MGP—the Mountain Gorilla Project, a cooperative conservation effort founded in 1979—had staff members working with gorilla groups that would be viewed by tourists. These groups were beginning to tolerate strangers, but weren't really calm about it. Neither were the rest of us. We would follow the rules to the letter: No closer than 15 feet to the gorillas, definitely no touching them. Nothing louder than a whisper. No visit longer than one hour.

Usually, when you're tracking mountain gorillas, the first thing you see is feces. Maybe you come upon night nests, and the trackers know you're on a fresh trail. Then you start seeing dung that's steaming. You know you're really close when you smell a musky sweetish odor. Some people find it repellent, but it's like perfume to me. The guides start making BVs—belch vocalizations, sounds gorillas make when they're content. It's not that you're pretending to be gorillas, it's just to say you don't mean any harm.

The first gorillas I visited were called Group 13, or Mrithi's group. They lived in a zone of bamboo near Mount Sabinyo. The bamboo forest is surreal and slightly claustrophobic because it's so dense and the light is really dim. Everybody said it was hard to habituate this group because you couldn't get close without being too close, and the gorillas couldn't see you very well. Also the group had lost its leader, a silverback shot by a poacher.

Mrithi was the obvious successor, but young for the role. He was at least 11; his back was turning from black to silver; and he seemed very unsure of himself. His mother, who seemed to be very old, gave me the impression of giving him a lot of advice. She would grunt at people a lot, the sound called pig-grunting. Mark Condiotti, of the MGP, said she wasn't happy about our being there, but he didn't take Mrithi's charges very seriously.

In a direct charge, a gorilla comes straight toward you, screaming. It's really frightening. Mrithi never did that. He would move sideways, and slam bamboo down and show his teeth, but you could tell he wasn't in earnest, just asserting his strength and showing he would protect the young ones.

One of the trackers, Big Nemeye, found a three-year-old gorilla, Mtoto, caught in a poacher's snare. Mrithi was charging around nearby, very upset. Nemeye cut the wire to set her free, and the next day Mark and two of the scientists went out to get the wire off and dart her with penicillin. They wouldn't let me go, but they told me that Mrithi never attacked them. He showed a lot of aggression, but he let them do their business. You can speculate about that and wonder what he was thinking.

Once, deep in the bamboo, Mrithi let me get fairly close; with a 300mm lens I could just frame his face. The light was so low I needed more than

Hunkering down, a silverback mountain gorilla known as Ziz waits out a chilly downpour, common in the Virunga rain forest. A brooding look suggests the weight of his responsibilities as leader of the famous Group 5, long studied by Dian Fossey. Maturing at about 12 or 13 years, blackback males gradually acquire the silver tinge of adulthood. Through intelligence and force of character, a dominant silverback will gain and maintain authority over his family group until he dies or yields position to another. A successful leader, Ziz has expanded his group from 10 to 34 individuals.

half-a-second exposure—and Mrithi moved, so his face is partly blurred. For me that was the only artistic picture I've ever made of a gorilla.

When gorillas are used to you, they glance at you often but they avert their eyes very quickly. You should do that too; it's very impolite to stare. I often had a huge lens in front of my face, but felt I was breaking the rules.

A lot of gorillas seem curious about the camera. Some, especially infants and young females, like to come up and look in the glass. One female wouldn't let me get a picture of anything but her; she was always sitting in front of the lens admiring herself in the reflection. I don't doubt that was what she was doing, because she was very much a princess. (You get a sense of personality very quickly with gorillas.)

———

Eating and traveling and resting are the events of a normal day. (This close to the Equator, day length doesn't change significantly.) After daybreak gorillas leave their nests and start eating. Between 10 a.m. and midday they settle down to rest or snooze. Then they rouse themselves to eat and travel until about 5 p.m., when they make their night nests and turn in for a good 11 hours. It takes a lot of plant food for those big bodies. On occasion Mrithi might make himself a tunnel in a wall of vegetation that you couldn't even see through: not eating it out, just barging through. The daily routine sometimes includes a bit of sex for the adults and always involves grooming as well as plenty of play for the babies and youngsters—running around, swinging on a vine, batting at twigs, tussling and chasing one another.

Sometimes it seems that a group likes to get up high on the volcanoes for the view. One day Mark Condiotti and I followed Mrithi's group straight from the nest up to 12,000 feet. They spent the day eating soil at a spot so little they had to take turns. Obviously it must have some mineral they need now and then. When they left this spot it was late afternoon. Females and youngsters would try to stop and feed whenever they passed some succulent food. Mrithi would come along and swat them: Give 'em a good punch, keep 'em moving. The females made a lot of mumbling, whimpering, complaining noises. We got down to a beautiful meadow, and Mrithi stopped, and they all pigged out till dusk on thistles and bamboo shoots at the edge of the forest. It was a classic case of silverback leadership.

From just 6 animals in 1980, Mrithi's group had grown to 14 by the time I went back in 1988. Every mature female had an infant. Tourists clearly hadn't disturbed the Rwandan gorillas' family life; but a mysterious respiratory disease killed 6 animals, and later on as a precaution 70 were inoculated against measles. The MGP and Rwanda's parks-and-tourist office had made the gorillas a great asset for the country, and checked poaching quite successfully, but everything about gorilla survival turns on complications in human affairs.

My colleague Tim Cahill compared our 1980 visit to a trip into Eden.

But this Eden is a tiny little island surrounded by overpopulation and hunger and despair. Not the human condition at its finest. Now, when you're with a mountain gorilla, you can often hear people talking and working in the *shambas*—the fields they farm. On my last visit, I heard bombs and guns.

The area where mountain gorillas live is divided among Zaire, Uganda, and Rwanda. There has long been trouble here between the Tutsi—the tall, cattle-owning warriors—and the Hutu farmers who used to be their subjects. In 1959 the Hutu in Rwanda launched a successful revolt and took control of the government. Recently, Tutsi refugees started a war against the Rwandan army—and the Parc National des Volcans became a battleground. In April 1991, I was with one of Dian Fossey's study groups, Group 5, on Mount Visoke. Machine guns and mortar fire came from the home range of Group 13. Both sides, however, had vowed not to hurt the gorillas.

A few weeks later I got a call from Diana McMeekin, of the African Wildlife Foundation, telling me that Mrithi had been shot. Newspapers around the world ran obituaries. Our friend had appeared in the movie *Gorillas in the Mist*, with Sigourney Weaver playing Fossey, and in television features as well. "He was the most reliable, reasonable, even magnanimous type of gorilla," Diana told the press. "He was utterly trustworthy."

Apparently soldiers disturbed his group in the dark, and he barked a warning from his nest. The fatal shots were fired in panic.

I had already come to expect that I would hear such news someday. I felt a special sadness for Mrithi, but more of the special sadness that I feel for the whole species. That they're under such pressure. That the world has them in that kind of vise. No visitor who's sensitive at all can miss the fact. You drive up a rough road; you walk through the fields and think, "Wow, there's a lot of people around here." You come to the forest, all of a sudden, and the guides say, "You must be quiet now." Before you know it you're sitting among gorillas. They're all around you, eating. Just living their lives.

With all the apes I've seen in the wild, there's nothing that compares to sitting in a glade and watching a family of mountain gorillas. They're beautiful creatures. They're so gentle. The silverback is dignified even when he's eating and picking his teeth, or lolling around passing gas, or letting infants use his back for a sliding board.

Whenever I have left the Virungas, I would make sure I would see the gorillas on my last day. Not to bang away so much with the camera, just to watch them. I would always have tears in my eyes before I left. I hoped I wasn't just taking from the gorillas, that I was giving something back, because they had let me into their world.

It's not like that with the western lowland gorillas, the gorillas that most of us know from zoos. The major populations known are in the intact lowland forests of Gabon, Congo, Cameroon, Equatorial Guinea, and the Central African Republic. They've been hunted for meat, for sorcery, for the pet and zoo trade. They're so wary they're almost impossible to study.

> *I hoped I wasn't just taking from the gorillas, that I was giving something back, because they had let me into their world.*

Since 1983, Caroline Tutin has had a research site in Gabon, at Lopé Wildlife Reserve. She had completed a research project with Jane Goodall at Gombe, and I remember the excitement in the primate community that Dr. Tutin would attempt the first study of chimpanzees and gorillas in the same habitat. How did two of man's closest relatives react to each other? It took almost a decade for Tutin and her colleague Michel Fernandez to get close enough to either species for consistent observations.

In the lowland forest, the apes can't see the humans at a comfortable distance and begin to get used to them. George Schaller, in the Virungas, had clearings and ravines in his favor; the gorillas could see him on the other side and not be frightened—they knew he couldn't fly.

———

J. Michael Fay and Richard Carroll did two-year baseline studies in the Central African Republic, near the border of Congo and Cameroon. Rich's site, Bai Hokou, is in a mosaic of uncut and selectively logged forest. Mike's site, Ndakan, a few miles away on the Sangha River, is uncut primary forest. Gorillas are thought to favor foraging in secondary vegetation, so these sites were ideal for testing the theory. At both, the gorillas favored secondary growth, but Ndakan—farther from human settlement—had about four gorillas per square kilometer in contrast to one at Bai Hokou. Both Rich and Mike did their doctoral studies with few direct observations. They studied distribution by looking at nesting sites. They could learn about gorilla diet by analyzing feces and feeding trails. They collected hair from abandoned nests for DNA studies. Once Mike followed a particular group for 17 days, staying a few hours behind so he wouldn't scare them and lose them.

In 1990, I spent seven days trying to photograph Bai Hokou gorillas with Melissa Remis, a Ph.D. candidate from Yale. She planned a thesis on positional behavior—how the animals move along the ground, how they move through the trees, how they use their hands—and was working under difficulties collecting her observations.

Every day we would go out with two Pygmy guides and have no luck at all. We never smelled gorillas. We never heard them. It's very difficult to track gorillas in lowland forest where there's little understory. In Rwanda the mountain gorillas made a path like a bulldozer's and even I could track them.

From what Mike had told me, his site sounded more promising. Unfortunately, before I got there I had fallen on a slippery trail and torn the cartilage in my left leg. I didn't want to say much about it, and Mike pushed us hard so I could get pictures. The local Africans call him Monsieur Béton— Mr. Concrete—and when he pushes hard it's almost frightening. We would leave camp before dawn, walk all day, get back at dusk.

Every day we saw gorilla sign, caught whiffs of scent, even saw glimpses, but never had a photo opportunity. The Pygmies kept telling us—

well, telling me—that we wouldn't have a chance unless we kept real quiet. We got quieter and quieter. They were carrying my big lens (400mm f2.8, a huge heavy thing) and a tripod and other stuff, and getting tired of it. My bad leg was hurting, so I was taking aspirin and leaving more things in camp every day. I always carry a flashlight, but one day I left my little headlamp behind even though it really weighs nothing. That day we actually found gorillas.

We were walking with the main tracker in front, then Mike, then me and a Pygmy behind me. He noticed a print that all the rest of us missed: one knuckle print on one leaf in the path. Five minutes later we were among a family of gorillas, up in the trees feeding.

The first one I saw clearly was a female, about 30 feet up a small tree that was swaying under her weight. A silverback screamed: sort of like a lion's roar at a higher pitch. He ran off, and I heard the others running through the bush. The female stayed put. Fear gave her a fit of diarrhea, which is what always happens with apes. She transferred to a big tree and went to the top of it. Then we saw a blackback male risk himself and go up to protect her.

That was midmorning, and we spent the whole day with them. Around three in the afternoon Mike said, "Look, we're like eight miles from camp, we should leave because we have no lights." The sunlight I had been working in was so bad I thought I didn't have anything good on film. "Mike," I said, "we've got to stay until the light gets lower and more angular." "O.K.," he said, but the Pygmies were clearly upset.

The female clapped her hands a lot. Mike had told me about that. It's a form of communication for the female to the silverback, apparently a distress signal, used because these gorillas often feed in trees where they can't see each other. We could hear the silverback close by, but he never came into view. I did get a good shot of the female peering down at me from 150 feet up in her tree; I used the 400mm lens with a 2x converter.

A beautiful sunset was visible in the upper canopy when we started walking home; our two subjects came down their trees and ran toward their group. Luckily we got back to our trail before dark—once the sun goes down in a rain forest, you can't see anything. It was Mike, two Pygmies, and myself holding on to each other, stumbling, falling over logs, landing in a creek.

The air wasn't too hot, around 70°F, but I was sweating, sweating from fear. I was sure I would put a hand on something that stung or bit. "We could sleep here in the forest," I suggested; "build a little fire." Nobody liked the idea. We had reached a bad section. We were crawling on our hands and knees because we couldn't stand up. Sometimes we did lose the trail for a bit. We made bad jokes and used a lot of profanity.

I got Mike to ask the guides what they did before the white men came along with flashlights. They said, "We were always home before dark." They got a big laugh out of that. My knee was killing me and I kept falling.

Finally we did build a fire. The Pygmies gathered twigs and tied them

together with vines to make torches—Pygmy flashlights. That's how we got home, shouting because we thought the camp staff would have sent out a search party. Well, no, when we came in they were all asleep, snoring.

"What's going wrong with the lowland gorilla studies?" I've heard critics raise the question. It's hard to say the terrain and the past are wrong, but they cause much of the trouble. I believe the Nouabalé-Ndoki forest in the Congo will turn out to be the breakthrough study site. Mike says it has an incredible density of gorillas and chimps, with no recent legacy of hunting. He is leading conservation organizations in an effort to persuade the Congolese to make it a national park. Imagine a forest where the apes do not grow up with the fear of man! Expeditions to date have seen the extrovert chimpanzees following out of curiosity and the shy gorillas a little uneasy but not frightened enough to leave.

━━━━━━━━━

One chimpanzee community has become quite indifferent to humans: at Gombe National Park. I went there, via Burundi, in December 1989, with Jane Goodall and a large group. It included her grown son, Hugo Eric Louis van Lawick (Grub), who was already a friend; one of her students; a five-member documentary film crew; an English conservationist; Burundi park guards; and an American diplomat—all helping with conservation work.

Gombe itself is no longer a camp of tents. Jane has a simple house, hidden among the trees by the lake. Her assistants and the staff of the park have huts. There are also huts for the tourists who stop, in erratic but substantial numbers, to see chimps. But here Jane is completely in her element. To see the chimps with her was a validating experience for me after a year following the disturbing path that chimps are taken on in captivity.

Following the Gombe chimps is very difficult physically. Their terrain is very hilly open scrub with thick riverine forest and areas of thorns and vines that are almost impenetrable to a six-foot human with cameras hanging from neck and shoulders. Everything grabs and snags. Most often when I emerged there were no chimps to be seen. When traveling they are short and close to the ground. They glide through the tangle.

Jane, in her fifties, wearing plastic sandals, can glide through the scrub as effortlessly as the chimps. She carries only her notebook and small camera. I would catch up with her torn, out of breath, and haggard. She would be perfectly composed. After one particularly hard climb she did comment, with a mischievous twinkle in her eye, "Someday I won't be able to do this."

Because of her work and Hugo van Lawick's photographs and films, I felt I knew the founding mother of a chimp clan, the famous Flo. Her daughter Fifi was carrying a young infant, Faustino. I made them my key subjects, hoping to show the strength of the bond between them. Fifi's adult sons, Freud and the younger Frodo, often traveled together; they would even let

Snacktime comes for a chimpanzee mother and her infant in the Taï Forest of West Africa's Côte d'Ivoire. The mother deftly wields a large stone to open nutshells without crushing the kernels within. She shares the food with her offspring, and, by example, her nutcracking techniques. Long-term studies of chimpanzees have revealed much variation from region to region in the use of tools.

me doze next to them after they had worn me out moving cross-country—when Frodo was not in a dangerous mood.

Ever since I began photographing apes, people have asked, "Aren't they dangerous? What is it like to be close to such powerful creatures?" Usually the apes are harmless. The only two exceptions I met were young males full of testosterone and used to being around humans from birth.

Pablo is a handsome mountain gorilla, the number two silverback in Group 5, one of Dian Fossey's famous study groups. As a precocious youngster he used to play with Dian. In 1988 the researchers warned me about him, but told me to stay close to the dominant silverback Ziz and then I wouldn't be bothered. One day I was shooting Pablo's portrait and didn't notice that the rest of the group had moved on. Suddenly the lens went black. Pablo calmly grabbed me by the shoulder and tossed me into the air. I rolled down the hillside, stunned and embarrassed, while Pablo followed the others. For the next month, I stayed close to Ziz.

Frodo, from what Jane says, has always been a bad kid. I was sitting next to her, taking his portrait, when he came up and grabbed me in a solid grip. I curled up in a fetal position and he spun me like a top. I feared he was building to a classic chimpanzee crescendo of aggression. Jane calmly snapped a picture; the click of her camera distracted Frodo, and he let me go. I went out of my way to avoid him after that.

Frodo's bullying techniques include rock-throwing, and he has made Jane a target in recent years. Gombe alumni, scientists who trained there, blame some of his bad behavior on tourism, which developed rather casually here. Frodo has seen tourists react—with laughter, yells, and flight—to his rampaging, and it encourages him. It inspires humans to plan another site for tourism, with a group habituated for the purpose.

Jane's project has filled in the cultural picture of chimps in comparatively open country at Gombe. In the Mahale Mountains, 90 miles to the south, Japanese biologists have been studying chimpanzees in more wooded country since the mid-1960s. A Swiss team in Côte d'Ivoire has found striking differences from both those populations. This study comes from tropical rain forest in Taï National Park.

Christophe Boesch and Hedwige Boesch-Achermann started their study in 1979, when there were few settlements nearby. The local people revered chimps and did not hunt them. Still, five years went by before a chimp group began to tolerate the observers—and even longer before the latter would allow a professional photographer on the scene. I was the first, in March 1991, and Christophe made it very clear that I should not disturb the research.

I shouldn't get closer to the animals than 20 feet, he said, unless they were too crowded and excited—as in making or sharing a kill—to notice.

*E*ver since I began photographing apes, people have asked, "Aren't they dangerous? What is it like to be close to such powerful creatures?"

That's how I got close-ups of screaming chimps tearing apart a red colobus monkey: not just dramatic shots but significant ones.

Taï chimps, says Christophe, share meat more than five times as often as those at Gombe, perhaps because cooperation is so important in the forest. "Here the success level goes down dramatically when one chimp hunts alone," he told me. In Taï National Park, 93 percent of recorded hunts involved two or more chimps, as against 36 percent at Gombe and only 24 percent at Mahale. The Boeschs regularly see some chimps drive the prey while others block an escape route and others wait in ambush. It happens in 63 percent of Taï hunts, only 8 percent of hunts at Gombe, and is unknown at Mahale.

I was happy to find the chimps' favorite nuts in season, because the use of stone tools for nutcracking is one of the Boeschs' major discoveries. Stones are rare in the forest, but chimpanzees in West Africa use them as anvils and as hammers. It takes about ten years to master the skill of breaking the nutshell without crushing the kernel, and mothers teach their infants carefully and patiently—cleaning the anvil stone, putting the nut in just the right position, demonstrating the proper grip.

Christophe is cautious; he's a scientist. But he does suggest that some human behaviors, such as hunting and long-term teaching, may have developed in a forest environment, not on the open savanna.

━━━━━━━━━━

If there are spirits of the forest, I think, they are in Borneo. That's what I felt when I first went there in 1983 to photograph the "persons of the forest"—the wild orangutans that survive only there and on the island of Sumatra. The trip was harrowing, with a flight in a weird plane like a bumblebee, a drive in a wrecked-jeep taxi, and a slow six-hour journey by boat to Biruté Galdikas' research site, Camp Leakey. The boat crossed a wide river, the Sungai Kumai, and turned into a small stream, the Sikunir Kanan.

I caught a glimpse of a small red-haired ape on the bank, a young one all alone and wailing pitifully. Otherwise we just went on forever and ever and ever, with the river getting smaller and the reeds and leaves coming closer, and it started getting dark. As I heard Biruté say recently, "You get to the camp at just about the time you think they're going to murder you and throw you in the river."

A long pier crossed a swampy ground between the river and the camp, and a man was dancing on it in the shadows. He turned out to be a Dayak, from the indigenous peoples of Borneo, dancing to Bruce Springsteen's music on his Walkman. Human life at the camp had a mix of cultures, and ape life was even more of a mix. Indonesian law forbids the keeping of orangs as pets, and several halfway houses try to prepare them for life in the wild. Biruté's camp was filled with these "rehabilitants," from bottle-fed infants to adolescents. Some of these youngsters would steal food and try to pull your clothes off, and they liked to pull film out of a cassette and wave it around.

Biruté was away when I arrived, but her Dayak observers were collecting data. Two men would follow an orang for five days, recording every action: everything it eats, every tree it visits, every time it defecates—anything they can see, especially an encounter with another orang. They trade watches, one resting in a hammock while the other works, and both constantly attacked by leeches, mosquitoes, and biting flies. They have to reach the nest before dawn and stay until the ape has built its new nest at nightfall.

At first the Dayaks refused to take me along, but I played volleyball with them and I pleaded a lot: "Please, my job depends on it, my life depends on it, I can't go home if I don't film one."

Finally they took me along to a big male called Sam. They were constantly pointing out something that they saw long before I did—flying frogs, flying lizards, a large black squirrel. When they found Sam, I could see his red hair in the trees. One day Sam encountered another male and came down to the ground to fight. The intruder took off running and Sam chased him and I chased both of them through the forest, sure I would get incredible photographs. Sam seemed determined to clobber the stranger. He looked over at me a few times and I thought, "Wooh, wait a minute—if he doesn't get the other guy, he'll probably just take it out on me." They have a really fierce look when they're worked up like that. Sam didn't catch his rival and I didn't get the pictures I hoped for, but it was a great experience.

The night before I left, one of the men, the quieter one of my forest partners, had some kind of seizure out on the pier. Some of the others told me that "evil forest spirits" had been attacking him. That night I seemed to see big male orangutans like Sam outside the window—silhouettes swaying on the trees. I kept the light on and stayed awake till dawn.

In Jakarta I met Biruté and told her about this incident. "Oh," she said, "he has epilepsy." "Sure," I agreed, "but what about the evil forest spirits?" She laughed softly: "That's just another thing in your life to forget."

Recently, however, I heard her lecture in Washington, D.C.: "The old Dayaks say that the forest spirits will be angry if we cut down all the ancient forest of Borneo." Some officials favor the logging industry over the forest, and without the forest, of course, the wild orangutan is doomed—it's the most arboreal of the great apes.

With ape photography, one of the things that disturbed me from the start was the fact that I was photographing arboreal creatures from the ground. I've climbed ropes most of my adult life, in deep caves and on mountain cliffs. I started thinking about ways to get up in the trees with my subjects. I ruled out chimps. I don't think you would be safe with them around while you hang on a rope. They might shake you up a bit. I chose the orangutan because it's solitary and kind of aloof. I discussed the idea with Biruté, and she said, "No, I don't think it's a good idea. It might be too upsetting to the orangutan because you'll be invading his or her domain."

I was still thinking about it in Sumatra. There the orangs are especially arboreal because of predators on the ground, not just man but also tigers. Indonesian Ph.D. students are working with the apes, following up on Biruté's work in Borneo. In June 1991, I reached Ketambe, a research center in northern Sumatra where Tatang Mitra Setia is studying wild orangs. Other scientists are studying the lesser apes, siamangs and white-handed gibbons, as well as monkeys.

Tatang was in the middle of a ten-day follow of Nur, the dominant male orangutan in the area. His students were following any females or other males that they could find. A minimum of ten days' data would show how the different orangs' movements and social activities interrelate.

Before dawn I was out with Tatang, walking through very hilly country to find Nur's nest. I had a fine example of a day studying an orangutan.

We reach the nest tree as Nur is waking up, stretching. He starts to eat. He eats different kinds of fruit, sometimes bark, a few leaves. He eats for about an hour, makes a day nest, takes a couple of hours' nap. He repeats this throughout the day. By late afternoon I have a crick in my neck from looking up all the time.

Suddenly Nur wakes up and starts a run through the trees, making an aggressive sound like smacking lips. He must think there's another male on his turf. He climbs the highest hill in the area. I'm not in good shape, but I manage to get ahead of him and shoot some interesting pictures—I run up the hillside so I can shoot straight across into a tree. Nur runs us until late afternoon. Up this hillside. Up, up, up, up. Finally he must satisfy himself that this other male has left; he heads back toward the tree he had been feeding in. Tatang tells me that Nur has been consorting with a female there.

It's dusk, and we want to hear limbs breaking and twigs snapping, proof that Nur's making a nest. "Please, Nur, go to sleep—you've had naps and we've been up for 13 hours." At last he starts nest building. Tatang is marking his map so we can find the site tomorrow. We hear a small panting sound within yards of us—that female is mating with another male. Nur screams. He's out of his nest and chasing his rival. Off we go again. It's around 7 p.m. before Nur makes a new night nest. We get to camp two hours later.

Next day a pattern develops. Nur leads us to a huge fig tree with ripe red fruit. He spends the next five days eating this fruit. His favored female is here too. She has a young one, eight or nine years old, old enough to be kicked out on its own if she gives birth again. That's what orangutan mothers do. They reject one offspring to care for the next, and some youngsters hang around for a time while others take off right away. This female is still being a good mother. But that's not the whole story.

A subadult male, Mr. X, is hanging around. When Nur's sleeping, or out of sight, the female leaves the fig tree and Mr. X promptly copulates with her. I watch this soap opera for several days and think about the plot. A subadult male is strong enough to take an adult female by force, and often

With ape photography, one of the things that disturbed me from the start was the fact that I was photographing arboreal creatures from the ground.

does, in spite of her screams and struggles. I don't notice any sign that Nur's female is resisting Mr. X. Maybe she just doesn't care. Females definitely seem to prefer dominant males as temporary consorts.

Mating aside, Nur seems intent on eating all the fruit in this fig tree. He makes day nests in the smaller trees nearby; he doesn't nest in the tree that is his food source. By now he has seen me a lot.

I ask Tatang about trying my plan for the trees. "Yeah," he says, "we'll try it and see. If it scares them, we won't do it. If there's any sign that they're really disturbed at all, we'll just stop."

I take the afternoon off and practice rope work with two young Indonesians. We take a crossbow and an arrow with a fishing line attached. Shoot the arrow over a tree limb. Recover the arrow, tie on a small cord, pull that down, tie in the climbing line, pull the rope down and tie it to a strong anchor. If you're going up 150 feet, you need at least 300 feet of rope.

Next morning we start shooting for a Y in a branch. All kinds of things go wrong. It takes most of the day to get the rope rigged. Late in the afternoon I climb the rope to my chosen branch.

There, at last, I really got a sense of what the forest is like for the orangutans. It's full of highways, with limbs and vines for trails in the lower canopy. Nur didn't pay any attention to me at all. He didn't seem any more surprised to see me up there than down on the ground.

I sat up there for three days, although it wasn't quite so good as I had expected. I hadn't allowed for the way limbs from my tree would hang in front of me. The orangs were hardly 30 feet away, and I would see little blotches of red every now and then. I didn't have any safety lines and didn't try to move around without them.

Still, I had a good time. I could see Nur go straight up the trunk of the fig. Whenever he went to sleep, Mr. X would come over to feed—he had a fine escape route 70 feet below me, so he could get away without Nur's catching him. Once I saw the female leave the fig tree and mate with Mr. X. Her young one wasn't big enough to reach the next tree on his own, so she made a bridge of her own body, holding vines and branches with hands and feet. She had to call her offspring, who was watching a strange creature— me. I saw wild siamangs approach the fig, making their unique haunting call. They were afraid to come in; I don't know if Nur deterred them or I did.

Perched in my tree, or chasing Nur and two other males until I was too sweaty to see them through the camera, I realized something. This is not my world. It's their world. They get around easily. When Nur chases another male, he comes down low to small trees that sway under his weight. He sways to another tree, bridges the gap, sways through the next. He really boogies through the forest that way. I'm just like a clown, dropping a lens in the mud, making too much noise, getting thorns in me. I can't take their world away. The things I've seen are better by far than the images I have. To convey it all, I would have to give you the smells and the sounds.

S *traddling the borders
of Zaire, Uganda, and Rwanda,
the volcanic Virungas rise nearly
15,000 feet, offering a montane
"island" for half of the world's
remaining mountain gorillas,
which number perhaps 600.
Before 1979, farmers encroached
on the area as they brought more
land under cultivation; since
then, Rwanda's government has
taken effective measures to protect
the park—a policy that helps
control erosion of farmlands
even as it preserves prime habitat.
Dian Fossey followed in the
footsteps of George B. Schaller
to study the gorillas here.*

*FOLLOWING PAGES: A silverback
known as Mrithi moves a bit—
and blurs his portrait into art.
Adult males can reach nearly 6
feet in height and 400 pounds in
weight, with an arm span of 8
feet. Adult females may weigh
200 pounds. Size and fierce mien
belie a largely gentle nature.*

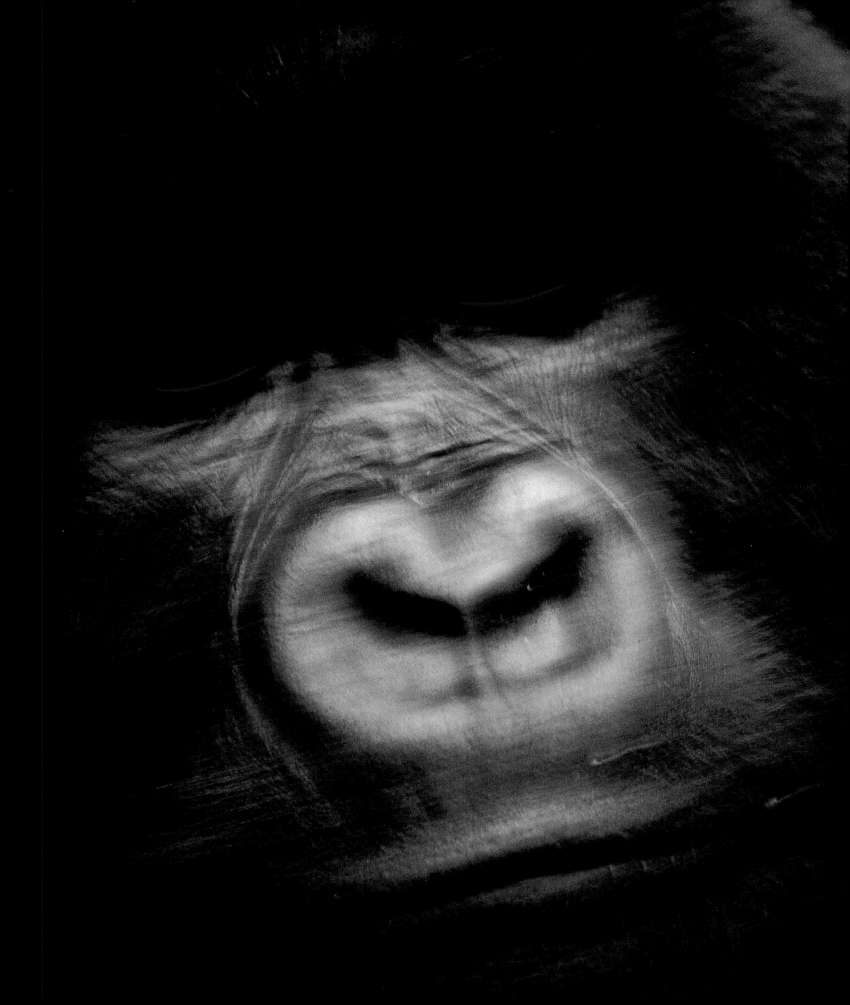

Sunnytime when the living is easy: Karisoke gorillas bask on a bright day, a luxury in the fogbound Virungas. Siestas punctuate days of feeding for silverback Beetsme and his family group. An infant suckles contentedly. Named—and misspelled—by Fossey in 1975 when he joined Group 4 as an adolescent stranger, Beetsme grew up to take control of the group, which had been disrupted by poachers.

Like "leisurely gourmets," gorillas "wander about, choosing a leaf here, a bit of bark there," wrote George Schaller. A dominant silverback, Cyane of Group 9, savors his choice of tender herbage. Almost exclusively vegetarian—though they may eat insects and other invertebrates—gorillas consume huge quantities to sustain their large bulk, stripping vines for leaves, chewing bark, fruits, flowers, berries, and sometimes roots of a wide variety of plants. The silverback determines where and when the group will forage, meandering perhaps a third of a mile a day before bedding down at dusk for the night. Sometimes, however, the leader will veer off unpredictably, hurrying straight to a distant site, probably in search of seasonal favorites such as bamboo shoots or a fruiting Pygeum africanum.

PRECEDING PAGES: Satiated after a morning's meal, apes of Group 4 snooze in the sun. "If gorillas had a religion, they would surely be sun worshippers," Schaller has said.

Papoose—first seen as an infant by Dian Fossey in 1967—cuddles her youngest offspring, Pasika. Diane Doran, director of the Karisoke Research Centre in 1989-91, quietly records behavior. A juvenile son, Kuryama, watches attentively while an adult daughter, sprawled at Doran's feet, pays no heed. Until 1991, when Rwanda's ongoing civil war first affected Karisoke, continuity of observation remained unbroken. "Gorillas are so long-lived," says Doran, "to understand the dynamics of groups takes a very long time." Two generations would require 16 to 20 years. Rain-drenched researcher Lorna Anness (above) hugs a soggy young gorilla in 1988. Once common, such intimate interaction has become rare. "The temptation is great," admits Doran, "especially with the little ones, but now we refrain from any contact." The change came to assure that research focused on gorilla behavior, she says, "not how they are with us." Now the war has interrupted research, and Karisoke's scientists have escaped—carrying computer records with them.

Mimicry links playful Peanuts and enraptured Dian Fossey in a Bob Campbell photograph, held up by a Karisoke worker. A moment after the picture was shot in 1970, the young blackback "extended his hand to touch his fingers against my own for a brief instant," she wrote. "The contact was among the most memorable of my life among the gorillas." That first gentle touch rewarded her patience in habituating her study group. "I tried to elicit their confidence and curiosity by acting like a gorilla," she explained. "I imitated their feeding and grooming"— as well as vocalizations. Grave markers at Karisoke bring to mind sadder times: Here lie gorillas fallen to poachers or to disease, including her favorite, Digit, slaughtered in 1977. "She seemed to have lost everything," Schaller wrote, "even despair," after Digit's death. In December 1985, she died, killed by an unknown assailant.

In a characteristic side charge, the silverback Mrithi crashes through brush, flailing stalks of bamboo. One researcher has likened such displays to "a symphony, beginning with slow hoots," with a "crashing finale," and said that afterward the animal "looked around as if expecting applause for his magnificent performance." For his efforts, Mrithi achieved a certain fame, in the movie Gorillas in the Mist and, more recently, as a star performer for tourists. Below, he searches for choice plants. Ecotourism has played an increasing role in protecting the gorillas and their habitat. Visitors would see certain groups, such as Mrithi's, carefully conditioned for exposure to tourists, leaving other groups isolated for continued research. Habituation has drawbacks: increased exposure to human diseases and aggression. In May 1992, in the midst of civil chaos, soldiers gunned down Mrithi before sunrise, leaving his 11-member family leaderless; it has disintegrated now. With the war continuing, other gorilla groups face similar threats.

Collared in clouds, gaunt ribs
of Mount Mikeno (the Naked One)
rise to 14,553 feet on the Zaire side
of the Virungas. A young gorilla
explores a garden of mosses,
lichens, and ferns sprouting along
a Hagenia branch. "Here the fairies
play," claimed museum collector
Carl Akeley, who died here in the
1920s and was buried on the slopes
of Mikeno. Akeley's efforts led to the
setting aside of the Belgian Congo's
Parc National Albert—divided after
1960 into Zaire's Parc National
des Virunga and Rwanda's Parc
National des Volcans.

Window opens into a new world on film for 1,500 Rwandans. In 1979, when the Mountain Gorilla Project launched an educational program, most Rwandans had never even seen a gorilla. This project aimed to teach the importance of the montane forest—in this desperately overcrowded land—for protection of watershed and prevention of erosion, and to show the benefits of ecotourism. A decade later, gorillas appear on the money—on Rwanda's 1,000-franc banknote— and figuratively in a boy's drawing of tourists with cameras snapping away at a group of happy gorillas.

The drawing held up reads: PARCS NATIONAUX (O·R·T·P·N GORILLE DE MONTAGNE

E xcept for birds and rainbows, the color here is green," says wildlife ecologist J. Michael Fay. In his dugout canoe he paddles down the Sangha River, border of Cameroon (the far shore) and the Central African Republic. He studies lowland gorillas nearby. A female, cornered on a branch 150 feet high, peers nervously down. She clapped her hands loudly, in typical female behavior, apparently to alert her group to her predicament. The lowland subspecies includes as many as 75,000 in the wild, as well as all gorillas in captivity. Few long-range studies, however, offer information on the wild populations. Dense undergrowth in the forest and extreme wariness—a result of long-time hunting by locals—make these apes difficult to observe. Fay now pursues his research in Nouabalé-Ndoki, a Congo region so inaccessible that hunting has not affected the animals' behavior.

Reaping reward for years of dedication, Jane Goodall beams at Gombe chimpanzees. "We're the longest study in the world—and, unless we stop, who can beat us?" she asks. During three decades she has gone far beyond early discoveries of tool use and meat consumption. The loss of a mother, she found, traumatizes a dependent infant far beyond the physical; but family members, even males, may nurture the bereft offspring. Orphaned in 1982 at age four, and now adolescent, Pax raises an arm to request grooming from his older brother, Prof. "That's an infant gesture," Goodall says, "which gets prolonged in orphans." Researchers in other parts of Africa have been working with wild chimpanzees nearly as long as Goodall. In Tanzania's Mahale Mountains, for instance, 90 miles south of Gombe, Japanese scientists have pursued field studies under the direction of Toshisada Nishida since 1965, observing behavior different from that seen at Gombe—particularly in matters of diet and tool use.

Family ties go deep at Gombe. Fifi relaxes with six-month-old son Faustino, while four-year-old daughter Flossi plays with a tree branch. Pivotal observations of the interactions between Flo—Fifi's mother, an old, high-ranking matriarch first seen in 1962—and her offspring revealed to researchers the influence maternal status and nurturing had on the young. Frodo, another of Fifi's sons (opposite), strikes a contemplative pose that belies an aggressive, rather bullying nature. Mothers and daughters remain close for life, Goodall found, and when stressed, adult males seek reassurance from their mothers. For her contribution to science, at her death in 1972, Flo rated an obituary in Britain's esteemed Sunday Times.

Brains, not brawn, advanced
Wilkie to the top at Gombe in 1990.
The alpha male peers cautiously
from his perch in a tangle of
branches. "He's not very big, but
that's not important. It's brains,"
Goodall says, "and he did have a
very aggressive, supportive mother,
Winkle." Alpha rank confers
privileges: freedom from attack by
other chimps; right-of-way on
paths; and first chance at favored
foods. Adolescent males begin their
climb to dominance by
intimidating adult females.
Timing and handling of social
relationships outweigh physical
aggression, according to Goodall:
"knowing when it's sensible
to display and when it's
sensible to retreat." Guile and the
strategic use of a shifting network
of allies—perhaps a brother or a
friend—further an aspiring male.
Evered, by family resemblance
possibly his biological father, helps
Wilkie maintain power.

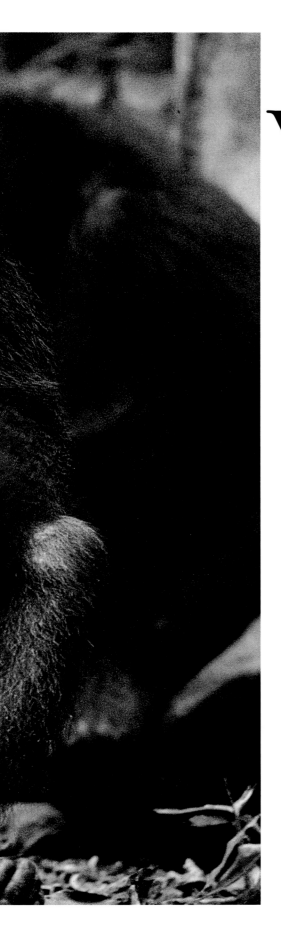

With stone hammer in hand, a mother chimp cracks nuts, an important part of her diet, while her infant quietly suckles. Observed since 1979, the chimpanzees of the Taï Forest of West Africa have demonstrated intriguing tool techniques not seen in East African communities. Chimps choose either wooden clubs or rocks—weighing as much as 45 pounds—to open nuts, using flat roots or stones for anvils. Indentations indicate repeated use of these anvils. Hard nuts, like panda, yield only to stones. For others, such as coula, wood suffices. Adult females exhibit greater dexterity than males in nutcracking and impart their skills to offspring. At age two, infants begin to consume nuts, and by age three or four they try to open them for themselves. "Mothers can stimulate nutcracking by leaving hammers or nuts near the anvil," says Dr. Christophe Boesch, a Swiss scientist who studies the Taï chimps.

Companionable chimps share
an African breadfruit, Treculia
africana, *in the Taï Forest.
A chimp clutches one of the large
fruits. "They seem to know
every tree and vine, and when
it is fruiting," says Christophe
Boesch, chief researcher of the
chimpanzee study that he and his
wife, Hedwige Boesch-Achermann,
initiated in 1979. The Boeschs have
discovered a tradition among Taï
chimps of sharing fruit and other
foods, which may have evolved
out of the mothers' offering nuts
to their young. Chimps reuse
nutcracking stones—scarce in the
rain forest—carrying them from
one tree to another and seeming
to have a "mental map" of where
they have left them.*

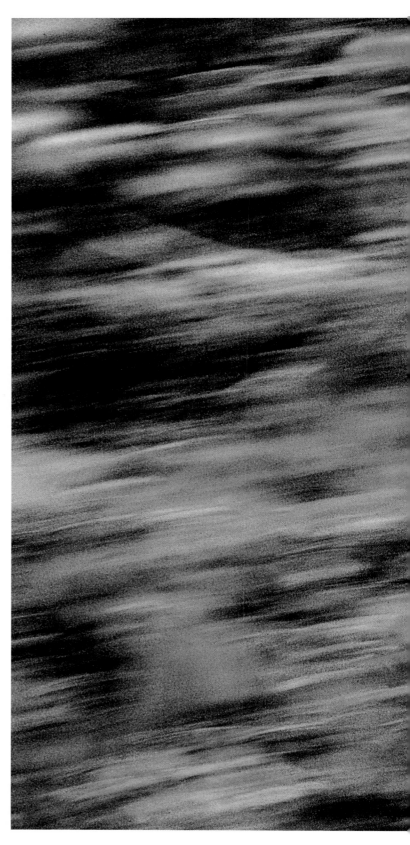

A dominant male, Brutus, scans
the forest canopy for red colobus
monkeys, the favorite prey of Taï
chimps. Another male races through
the forest in a drumming display:
He will leap and repeatedly smack
feet and hands against the buttress
roots of large trees to communicate
his location and direction of travel
to chimps scattered in the forest.

FOLLOWING PAGES: Brutus sits down
with the prize of the day, a hefty
colobus. The hunters can usually
find a red colobus weighing almost
30 pounds within about 20 minutes.

Frenzied chimps divide a red colobus, meat enough for the team. A fortunate female (above) chews leaves mixed with meat from the neck and head, presented to her by a male. In the dense Taï Forest, hunting requires tactics different from those observed in East Africa. Chimp "pushers" track the colobus and drive it toward "blockers," who anticipate the monkey's path and grab it. "What intrigued us," Boesch says, "was the degree to which they hunt cooperatively."

With a vine for an easy chair, a chimpanzee mother and her infant linger in the forest maze. Encompassing some 1,600 square miles in southwest Côte d'Ivoire, the Taï Forest shelters a rich range of animal life that includes more than 200 species of birds, 10 species of monkeys, 6 species of duikers, bushpigs, bongos, pygmy hippopotamuses, and forest elephants. The density of the forest makes these chimpanzees difficult to observe and may explain why researchers find marked differences between their behavior and that observed elsewhere. With visibility often limited to 65 feet, successful hunting requires cooperation and subsequent sharing of prey. Boesch has observed 19 variations of tool use and 6 ways of fashioning tools, compared with 16 kinds at Gombe and 3 methods of manufacture.

With teeth bared, a female Taï chimp defends the corpse of her infant. The first baby of this inexperienced mother fell from a limb, probably breaking its neck and setting the troop into shrieking pandemonium. For two days the mother carried the body around before finally giving it a gentle touch (opposite) and leaving it behind. In such behavior, researchers find the emotional responses of the apes akin to human feelings. Elements of labor division, teaching the young, and sharing food, demonstrated by Taï chimps, hold striking implications. "We hoped," says Boesch, "that what we learned about the behavior of forest chimpanzees would shed new light on prevailing theories of human evolution." The Taï studies, he suggests, may refute an accepted theory that conditions in the open savanna forced change upon early hominids. The forest presented more of a challenge, he asserts. "If the chimps of Taï are a model of how hunting strategies develop, then group hunting, cooperation, and the sharing of food could have evolved during the period when hominids dwelt in the forest, and not, as we used to think, on the plains."

Gentlest and most sensual of apes, bonobos inhabit a forest area south of the Zaire River. Here, Kiku cradles her firstborn son, Kikuo. Sons stay with the mother's group for life, while daughters join another. Females, whether of high status or low, may rank as equals with males. Sex, in all sorts of combinations, serves to reduce tensions in any exciting situation—finding a new food supply or meeting another group. An immature female and infant male (below), captives at Kinshasa, mimic copulation. Early in 1991, Takayoshi Kano observes from a hut in a clearing provisioned with sugarcane to bring bonobos into view. After years of study, he recognizes a hundred individuals in three groups.

FRANS LANTING (ALL)

Schoolchildren watch a bonobo male and a female carrying two young cross a rural road; they know the apes as bilia *(plural; elia is the singular).* Bonobos knuckle-walk on all fours but can go upright for short stretches. Last recognized and least studied of the great apes, they have come under increasing pressure from a fast-burgeoning human population, losing habitat to slash-and-burn agriculture and to logging interests. At most, 20,000 remain in the wild, and there are not enough of them in captivity to sustain a genetically viable breeding program. Now political instability clouds their prospects in Zaire. Foreign researchers had to leave the field in September 1991. A year later Dr. Kano paid a brief visit. He reports that local officials and chiefs promise to prevent poaching by outside hunters; trackers will follow his groups every day but Sunday; missionaries will mail the trackers' reports to Japan.

S *eventy feet above the forest floor, a subadult male orangutan called Mr. X uses vines to swing from tree to tree at his best speed. He's afraid because a dominant male named Nur is chasing him, says Tatang Mitra Setia, researcher at the Ketambe study area in northern Sumatra's Gunung Leuser National Park. In his teens the male develops the marks of maturity: cheek flanges, or pads; a beard and long body hair; and a throat pouch. In his prime he utters the famous "long call," a vibrant low rumble that rises to a roar and ebbs away. Biruté Galdikas, who for two decades has studied the Borneo subspecies, describes this call as "the most impressive and intimidating sound" of the forest; she thinks it not only warns off lesser males but also attracts estrous females. Except for consort pairs and mothers with young, orangutans usually forage alone. In her 14-square-mile study area, Galdikas has documented their use of 400 different foods, mostly plant stuff. "Orangutans are not built for speed," she says; "they're built for hanging from branches and eating."*

Under mother's watchful eye, an infant orangutan slurps milk dripped by a worker at Sumatra's Bohorok Rehabilitation Centre. After successful reintroduction into the wild, the female has brought her baby to the center's feeding platform. A younger, less self-assured rehabilitant called Yenny goes fearfully into the forest with ranger Supandri. Such halfway facilities complement a longtime ban on owning wild orangutans.

FOLLOWING PAGES: At Camp Leakey, Galdikas' rehabilitation center in Borneo's swamp forest, camp assistant Achyar brings sugarcane and manioc to rehabilitants. One young ape plays with a long-tailed macaque, a visitor from the wild.

Call of the wild at Camp Leakey: In hot pursuit of a wild adult male (above), ex-captive female Rinnie scales a tree (opposite). Wild adolescent females also court fully mature males, who seem more attracted to receptive adults as mates. Logging, mining, and other human activities now threaten the forest on which the orangutans' survival depends.

FOLLOWING PAGES: Rehabilitant mother Akmad broadens her family circle to include a young ex-captive female who sought her care.

U*nlikely places, unnatural activities, and unforeseen chances mark lives of the great apes in the human sphere, a fast-changing realm of incongruities. At the Stardust Resort & Casino in Las Vegas, Nevada, a young female orangutan named Bo appeared in 1990 in a show called "Lido de Paris." Wheelchair and socks protected her feet from harsh cleansers or other chemical residues.*

By Michael Nichols

The World of Man

I t was my first contact and, like yours, it was not with an ape in his natural home. I was in Montgomery, Alabama, in 1976 to visit my father, who by divorce seemed like a stranger. I stopped at the small zoo to relieve the tension of the pending meeting, our first since I had come out of the army. I found myself in front of a group of chimpanzees, the first apes I had seen outside the pages of NATIONAL GEOGRAPHIC. I was the only spectator, and the five animals concentrated their antics on me. Somersaults, hand-slapping, and tickling one another were just part of their repertoire. I felt very special at this one-act play, and stayed until closing time.

I took many photographs, which became a feature in my college newspaper. The chimps' facial expressions were used to illustrate the stress of exams. The faces all stared out from behind bars—something I cannot remember having noticed at the zoo.

For reasons I will not try to explain, humankind find it very handy to use apes as surrogates.

This must be one reason why controversy flourishes over the varied situations of apes in the world of man. Zookeepers, entertainers, and pet owners see themselves as friends of endangered species, helping to save them from extinction. But, say critics, their activities keep smugglers in business—and captivity at its best amounts to abuse. An animal dealer in West Africa told me that chimps were not endangered and that if he didn't export them they would just "end up in some villager's pepper soup." American biomedical researchers say they cannot study AIDS and hepatitis adequately without using chimpanzees; they don't publicize the lives of their subjects. A number of groups campaign for animal welfare or animal rights, with tactics that may include lawsuits or lawbreaking, and a few individuals simply devote themselves to apes in their care.

Late in 1989 I found myself again watching chimps making faces, hand-slapping, and razzing me. These were youngsters at the Chimfunshi Wildlife Orphanage on the 10,000-acre farm of David and Sheila Siddle in northern Zambia, near the Zaire border, which is now well to the south of chimp habitat. Customs officers would bring confiscated orphans to the farm, where David (who is nothing if not a dreamer) had recently built the Great Wall of Zambia. A high wall made of bricks fabricated on the farm, this enclosed 7 riverside acres of chimp playground. David was planning a larger sanctuary, about 2,000 acres, with electrified wire to be powered by solar panels. Such wire now surrounds a 14-acre habitat, housing a second group of chimps. At first the Siddles had paid for the project from the profits of their farm, but I understand that these days they do receive donations. At last count they had 43 orphans, two from a circus that went broke in Papua New Guinea.

Once removed from the wild, chimpanzees can never be returned to a place that has any chimp populations. The residents will attack and kill the invaders. Sanctuaries like the Siddles' or the Jane Goodall Institute's may serve

The baby face in a tiny window reflects the worldwide plight of captive chimpanzees. This one lived at a private zoo in Monrovia, Liberia: a rejected pet, or seized from hunters seeking to smuggle it into illegal trade. "Taken for profit by wildlife dealers," write Jane Goodall and Geza Teleki, "prized as subjects in medical research...valued as exhibits by zoos...advertised as attractions by the tourist trade, trained as performers in...entertainment, coveted as substitute children by pet owners, chimpanzees are trapped in their...close kinship with humanity." After civil war broke out in 1989, the Monrovia zoo animals were taken for food.

as educational centers, but some biologists think all efforts and funds should be devoted to the protection of habitat. Former captives can never be truly wild, and may be dangerous to humans as they have lost the fear of man and know their own strength compared to his.

In the United States and abroad, zoos have attempted to replicate natural environments. They emphasize exhibit areas without bars, education in nature's complexity, and propagation of endangered species with proper regard for genetic diversity. This trend has been especially good for the lowland gorillas. The New York Zoological Society—recently renamed NYZS/The Wildlife Conservation Society—is planning a new habitat that will take advantage of Mike Fay's current research, to display the gorillas' tree-climbing and fruit-eating behavior. Heated outdoor areas will extend the display season to about nine months, but the animals will spend the coldest weather indoors.

I have followed the outdoor liberation of two wild-born silverbacks: Zoo Atlanta's Willie B., who was confined alone behind bars for 27 years, and Timmy, captured as an infant, who spent much of his life alone at the Cleveland Metroparks Zoo. The American Association of Zoological Parks and Aquariums (AAZPA) has developed a "species survival plan" for breeding management of endangered species, including more than 300 lowland gorillas in the U.S., Canada, and Mexico. Officials decided to move Timmy to the Bronx Zoo (now designated as a wildlife conservation park) to join several fertile females. Many captives have a loyal following of "gorilla groupies"— Timmy's fans in Cleveland took pride in the move. He was sent to the Bronx and carefully made acquainted over the winter with three possible mates, Pattycake, Tunko, and Julia.

I watched through a 600mm lens in the spring of 1992 as Timmy had some of his first moments outdoors since his capture. The females charged out into the exhibit space while Timmy peered out warily from a doorway. After a day or so I saw him move to a large rock by the door and nervously take in this new world around him, flinching as birds or the zoo's aerial tram passed above him. A great moment came when he let his huge hands and feet touch the grass, keeping one foot in contact with the rock and quickly sniffing the other. After one week his obvious nervousness was gone. "Timmy & Pattycake, sittin' in a tree, K-I-S-S-I-N-G . . ." announced the New York *Daily News*, and at last report Pattycake was pregnant.

The two thousand chimpanzees used by biomedical labs in the United States do not get headlines. Their service to humankind goes without glory. Maybe if a vaccine or cure for AIDS is announced, a chimpanzee will be *Time*'s "Man of the Year" instead of the scientist who used it. The virus infects chimpanzees, but so far has never made one ill; this makes them the prime choice for experiments we do not subject humans to.

This subject troubles me very much. Of course I want to see diseases defeated; but I do not believe that it is man's inalienable right to do with the earth and its creatures as he pleases. There is a genuine quandary here. We invoke "bad guys" to explain such things. Unfortunately, if there're bad guys they did not let me into their world to photograph.

Because I entered the biomedical community under the auspices of Dr. Goodall, I dealt only with labs that were sympathetic to her proposed changes or felt it safe to admit a journalist. In two years I was allowed to work in three labs: LEMSIP, 45 miles northwest of New York City; the Southwest Foundation for Biomedical Research in San Antonio, Texas; and Vilab II in Liberia. Rejections were too numerous to mention. In one case I was turned down even though I was to accompany a U.S. Senator on a visit to a facility supported by federal funding.

LEMSIP (Laboratory for Experimental Medicine and Surgery in Primates) holds about 250 chimpanzees used in all types of experiments, with AIDS studies a high priority. There is no outdoor holding or exercise area, but the hall for the HIV chimps is designed for state-of-the-art arrangements with roomy cages to maintain their physical health.

Entering that unit for the first time, on a chilly October day, was a disturbing business. In a dressing room like an air lock, I put on a thin white disposable coverall, booties, a surgical face mask, hair cover, and latex gloves. I'm not sure whether I was protecting myself or the chimps. I stepped into the tropical warmth of the hall and a dozen chimpanzees greeted me with a deafening chorus of banging and hooting. They had laid claim to this dismal territory and defended it by spitting on me. They also threw feces, a trait that has made captive chimps the least popular of the great apes with many keepers. I was sweating, my cameras fogged up, and that visit was useful only in the lessons it taught.

The caretakers at the labs I visited are often dedicated animal lovers. They believe their job is to take the best possible care of their subjects under the circumstances. The chief veterinarian at LEMSIP, Dr. James Mahoney, is a tortured soul. He genuinely loves "his" chimps, but his job is to produce as many young ones as research will need. He romps and roughhouses with the infants in the nursery. Volunteers pamper the nursery chimps. The regular staff gave the impression of holding emotion in check—they were professionals on the job—but they will take a sick infant home overnight or do whatever else is needed when affection must supplement skill.

Dennis Helmling, a husbandry technician when I met him, asked for the job of working with the AIDS chimps. I asked why he chose such a difficult, thankless task. He said quietly, "It has to be done and I'm happy to do it." He would talk to each chimp as he cleaned its cage. He tried "behavioral enrichment" items: combs, toothbrushes, mirrors, bubble-blowing devices, tapes of music, TV. Billy Jo, a former circus performer, would stand on his head for Dennis (labs are a favorite dumping ground for animals too mature to be controllable in the entertainment industry).

For reasons I will not try to explain, humankind find it very handy to use apes as surrogates.

Two years later I was back at LEMSIP and I acclimatized my cameras this time. When I walked in with Dennis, the chimps barely acknowledged my presence. His attentions had paid off. When he gave out Kool-Aid, a favorite treat, each chimp patiently waited its turn. I said, "They sure are calmer now!" He just smiled.

Long-term care of infected chimps is a major concern. Dr. Jorg Eichberg, formerly in charge of the colony at the Southwest Foundation, developed a "retirement" plan. The pharmaceutical company funding the research must set aside money for the chimp's care until its death. Unfortunately, the retirement quarters I saw in 1990 were stark cinder block housing in a small yard. Our imagination often seems severely limited on this score.

Primarily Primates, Inc., a private sanctuary near San Antonio, houses unwanted survivors from research programs, zoos, the pet trade, and entertainment enterprises. There I saw Tyrone the Terrible, a chimp rescued by animal lovers from a small circus, and Joe, an old chimp that a ramshackle billboard once called the world's largest and most ferocious gorilla. Joe had spent years in a small filthy cage, its door rusted shut; when he was freed, his muscles were so atrophied he could hardly walk. By 1990 his gait had become normal again, but he hoarded food like a former prisoner of war. Primarily Primates, not open to the public, depends entirely on donations to support 325 apes, monkeys, and lemurs; each cage and enclosure has a plaque honoring an individual donor.

Donations are luxuries for Sam, an adult chimp living outside a bar at Maineville, Ohio. His cage is surrounded by signs. These say that he has a heated underground "condo" with a TV set; that he enjoys action films and soap operas; and that he likes to smoke and drink but shouldn't be given anything. During my short stay, a few visitors brought healthful items like grapes and sliced apples, which he ignored. Others gave him seven soft drinks, mostly colas; lots of cheesy snacks; and cigarettes, which he lighted with his own lighter. After the jolts of caffeine he did some classic male chimp displays. Sam and his owner, I was told, have an emotional investment in each other; a scientist who used to teach nearby said, "Sam really gets a healthy diet—and at least he has a relationship with somebody."

Pet owners, in my experience, have intense, almost obsessive relationships with their apes, and are troubled by the knowledge that Jane Goodall doesn't approve of such situations. "If I could just talk to her!" said one. "I don't make them perform, and I go around to schools giving talks about how endangered they are!"

I heard the famous entertainer Bobby Berosini discussing the theme of endangered species at a public school in Las Vegas; his star, the young female orangutan Bo, appeared with him. For eight years he and his orangutans performed at the Stardust Resort & Casino, in an adults-only show. I saw the act about ten times and never was it the same; the variable was not

Recent graduates of nursery ritual observe the changing scene at the San Diego Wild Animal Park, a sister facility of the San Diego Zoo. In such sanctuaries most humans find their only opportunity to come face-to-face with other primates. This preserve offers semblances of natural habitat for flocks and herds, for families of primates to live and breed. And a nursery too, for baby bonobo Ikela, when her mother neglected maternal chores. Restricted to one forest region of strife-torn Zaire, pressured by loggers, farmers, and poachers, bonobos—or pygmy chimpanzees— number at most only 20,000. San Diego and other zoos are expanding their roles in research and breeding programs to ensure survival of such endangered species.

Bobby but the reactions of the orangutans to the occasion. But times change, and his show has moved to Branson, Missouri.

Come to think of it, all of the "entertainment apes" that I've photographed are from a bygone era. The oldest is Jiggie, now in his late fifties; he played the chimpanzee sidekick Cheeta to Johnny Weissmuller's classic Tarzan in the black-and-white movies of the 1940s. The notable exception is a human who has trained himself for other primate roles: the English actor Peter Elliott, who played two ape characters in *Gorillas in the Mist.* Even without a costume, he can evoke a hooting, swaggering chimpanzee — as I once saw him do in a pub.

In a category of one is Kanzi, the bonobo prodigy who lives and studies under the tutelage of behavioral biologist Sue Savage-Rumbaugh at the Language Research Center near Atlanta. For two years Sue tried to teach his wild-caught mother, Matata, the use of colorful symbols called lexigrams as meaningful words. Kanzi, from six months on, hung around and watched; now and then he used the keyboard of lexigrams. When Matata was taken away for breeding, Kanzi suddenly displayed the skill his mother had never learned. On the first day of her absence, he produced 120 separate utterances, using 12 different symbols. "I was in a state of disbelief," Sue says; she "had not intentionally taught Kanzi anything." He had learned on his own — as wild infants learn by watching and imitating their elders.

It is hard indeed to resist the spell of this captivating animal. Kanzi can understand spoken English very well. His comprehension far exceeds his ability to "speak" with the keyboard and computer at his disposal. Now, as an adult, he can be aggressive on meeting human males and, like all apes, he reads body language very well. Even with a glass wall between us, I was careful to move slowly, keep my arms and hands close to my body, avert my eyes, and be as self-contained as possible. In the wild, it's easier to remember to go by "their rules" — with captives, you're usually being bombarded by the ways of your own world.

Kanzi is truly an ape between two worlds. He still feels the drives and instincts that make him a bonobo, but he has been able to apply his intellectual energy to understand and use a "foreign language" and communicate with his human guides. Sue continues to publish her findings for the research community, which is taking renewed interest in such work, and suggests that the great apes have "a greater propensity for the acquisition of symbols" than anyone recognized in the past.

Much as I admire what has been accomplished with captives, I find even more valuable Christophe Boesch's goal of understanding ape intelligence through the way his subjects deal with their natural world. If we can learn to translate their "foreign language," I believe the depth of its complexity will humble us. Yet once in the world of man, that is where most captives stay. I believe that we must evolve to accept them for what they are and treat them by standards we currently reserve for ourselves alone.

Panbanisha, a young bonobo, lets her fingers do the talking on a plastic-covered sheet. Her forefinger rests on a white square with a letter Y, meaning "yes," to answer a question about a favorite destination. Called lexigrams, the symbols in the array signify things, places, individuals, feelings, and actions. Color does not code for meaning, but improves readability. Apes cannot make the varied consonant sounds that distinguish human speech — can they learn to communicate with sign language or symbols? At Georgia State University's Language Research Center, near Atlanta, scientists explore the mental processes of acquiring knowledge and the ability of Panbanisha and others — of both chimpanzee species — to develop language skills with lexigrams.

S ocial climbers sound off at a chimp klatch in Zambia, where David and Sheila Siddle have converted their farm into the Chimfunshi Wildlife Orphanage. Big Jane perches above open-mouthed Boo Boo, dark-faced Little Jane, and Coco. Above, Tara leads, while the Siddles carry Rita and Sandy; head keeper Patrick Chambatu walks behind. From aggressive, mentally troubled chimp orphans, David and Sheila struggle to knit stable groups with a recognized hierarchy, the natural lifeway of the species. Chimfunshi, says Michael Nichols, is "a story of real heroes in the middle of nowhere in Africa."

FOLLOWING PAGES: Loners in the wild, orphaned young orangutans draw together in the trauma of captivity. Among ten smuggled from Indonesia for Taiwan pet fanciers, they were intercepted and sent back to Jakarta. Return of the Taiwan Ten marks only a partial victory — two of the ten have died; three remain in an Indonesian medical lab.

Congolese staff provide daily field trips—practice in climbing and foraging—in woodland around the Brazzaville Gorilla Orphanage: Albertine N'Dokila tends Dinga, Kola, and Billinga. Confiscated from poachers and illegal traders, sick and wounded little ones recover and grow here. A government reserve with special protection is planned for them. Before the orphanage opened in 1989, with funding from British conservationist John Aspinall, Yvette Leroy of Brazzaville undertook to care for gorilla foundlings—all too numerous now.

Ape meat, ape magic, and the ape trade lure hunters into the bush. From night stalking with headlamps—demonstrated here by a Liberian hunter—chimpanzees have virtually no escape, says Dr. Geza Teleki, director of the Committee for Conservation and Care of Chimpanzees. Once, hunters took game for village cook pots; now, writes Teleki, they "supply bushmeat in volume to urban markets." Once, hunters shunned adult females. Now, he estimates that at least ten chimpanzees die for every baby received by an overseas buyer. Gorilla hands, on sale illegally in a Congo fetish stall, are thought to transfer the great ape's power to humans. One recipe: Pulverize a fingerbone, rub the powder into a cut on your arm.

Solitary confinement—withering
fate for a chimpanzee, destined by
nature to live gregariously—chains
young Whiskey inside a garage
in Bujumbura, capital of Burundi.
Like many an orphaned chimp,
he became a cute young pet for
a foreign resident in Africa. Like
most pet chimps, growing into the
strength of maturity, he became
uncute and unmanageable. For
Whiskey, release came through the
Burundi Chimpanzee Conservation
Program, affiliated with the Jane
Goodall Institute. At the program's
"halfway house" in Bujumbura,
he received careful nurture,
with a rekindling of social ties;
but he already had tuberculosis
and it claimed his life in 1991.
For luckier chimps, JGI envisions
a Burundi sanctuary where
rehabilitants will roam in lively
groups, a source of knowledge
for scientists, a source of
delight for tourists.

For sale in the Kasbah of Las Palmas: your picture with a chimpanzee, sad-faced though gaily clad. Photographers ply the Canary Island beach resorts, hauling the young chimps about in sultry sun and smoky discos. Conservationists trace the supply line of chimps to smugglers in Equatorial Guinea, a former Spanish colony. To Jane Goodall, the apes seem listless, drugged; and so they appeared to her son, Hugo van Lawick (opposite), holding the chimp in red next to the apes' owner. Goodall uttered a few pant-grunts of greeting. A beach photographer, now suspicious, reached out for the chimp, which clung to Goodall, reluctant to let go of a newfound friend. "It was terrible to have to leave them," she says.

FOLLOWING PAGES: On a happier day she chimp-hoots to a dancer mimicking the pant-grunt and swagger— celebrating the founding of a sanctuary near Pointe Noire, Congo, by Conoco Inc. and the Jane Goodall Institute.

Broad-beamed Willie B., the lowland gorilla peering through the greenery of Zoo Atlanta, survived a cramped, boring cage to relish his lush new digs at the Ford African Rainforest. For 27 years the "loneliest gorilla in the world" had never seen such flora—or even another of his kind. Now he shared a sampling of primeval habitat, adapted to survive freezing and foraging, with 13 others on loan from the Yerkes Primate Research Center of Emory University. Zoogoers approved. So did the gorillas; within a year their tribe increased by three. A visitor studies an orangutan (below) at the Pittsburgh Zoo, which simulates natural settings for 16 endangered primate species.

FOLLOWING PAGES: *Specters in a foggy realm, lowland gorillas (right) haunt Myombe Reserve at Busch Gardens in Tampa. Given such a setting, humans may appreciate great apes as more than gymnasts, clowns, or monsters.*

R and R at the Great Ape House: Mokolo the lowland gorilla rests on his bed of straw, while Batu the orangutan (above) works the ropes at Chicago's Lincoln Park Zoo. Though it rose in 1976, before naturalism replaced straw with misty verdure, the house mingled structural innovation and insights of science to produce a world-famous primate program. Ropes simulate jungle vines; poles make do for trees. Food strewn in the straw stimulates "foraging." From pioneering studies of George Schaller and Dian Fossey, keepers learned the truths of gorilla society and formed family groups ruled by silverbacks. These groups are producing healthy young.

Pampered, but not pets, the easy riders of Liberia enjoy their version of a pram at a biomedical facility. The species plays a prominent role in the quest for vaccines against hepatitis, AIDS, and river blindness, one of the scourges of tropical Africa; but a treaty curbs exports from Africa and U.S. law restricts imports. In 1975 the New York Blood Center established a research station, Vilab II, near Monrovia, and it founded breeding colonies. Animal welfare groups often protest that researchers ignore their subjects' psychosocial needs. Vilab II kept young with mothers; if mothers rejected infants, staff took over. Caretakers, such as Jennah Dolley transporting her charges here, spent all day with the two-year-olds, playing with them, letting them climb trees. In their post-research days, chimp groups would romp freely on retirement islands. War in Liberia has brought tragedy—looting, and the murder of Vilab official Brian Garnham. Jennah Dolley still cares for two infants known to survive.

Gravely ill, a foundling chimp clutches at foster comfort. His mother vanished after giving birth on a Vilab II retirement island; caretakers discovered the ailing baby with two males trying in vain to play stepmother. He died a few days after this photograph was made. Many captive mothers themselves lack care-giving skills. A mother who cares (opposite) lends a comforting hand to her bemused offspring; Vilab encouraged such bonding. Labs exempt young chimps from experimental regimens until they reach three to five years. To allay boredom, Vilab animals lived in groups. Adults had the chance to tinker with tools— sticks to dig jam or peanut butter out of jars set outside the cages, and stones to crack palm nuts delivered in the shell.

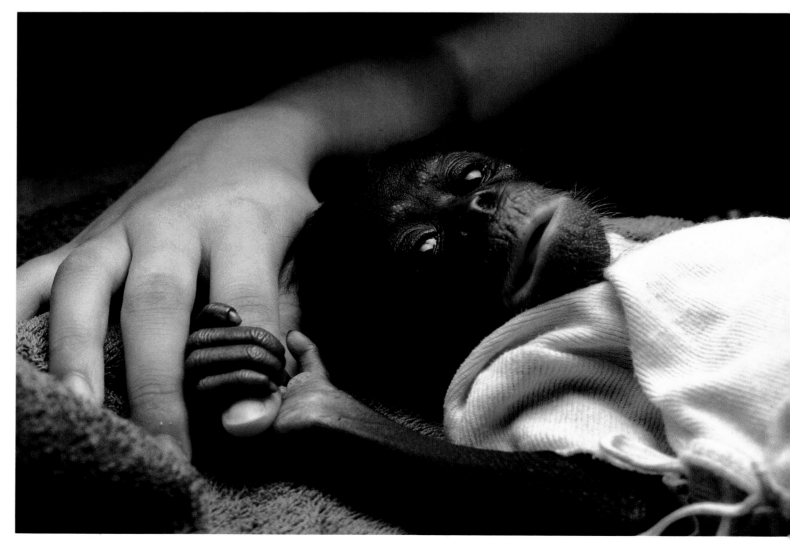

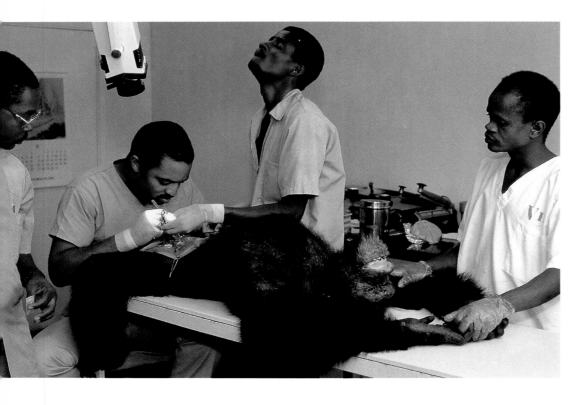

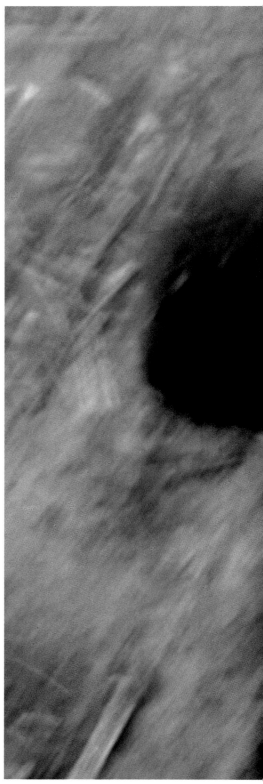

A surgical team performs a
sterilization at Vilab II—to slow
chimp breeding that produced a
bumper crop, in excess of demand.
Vilab turned to birth control even
as American researchers worried
about shortages and chimpanzees
in the wild continued their decline.
Why not restock forests with
captive-bred chimps? They lack
the social skills and habits that
give chimp society its strength,
and wild ones attack them as
intruders. On Vilab's retirement
islands, caretakers delivered
supplementary food. In 1990, when
the Liberian war cut access to the
islands, 49 Vilab chimps died—
starvation or poaching the most

likely causes. Right: An adult male
at Vilab vents rage and frustration,
a common outbreak among grown
male chimps in confinement and
a source of displays in the wild.

FOLLOWING PAGES: A cooling
squirt from friendly technician
Dennis Helmling relieves cage
monotony for an HIV-infected
chimp at the LEMSIP laboratory
in New York State. Mask and
gloves protect against virus-
tainted blood, saliva, and feces.
Because AIDS cannot be
transmitted through the air,
nobody put filters in
his respirator, which is worn
simply to shield his face.

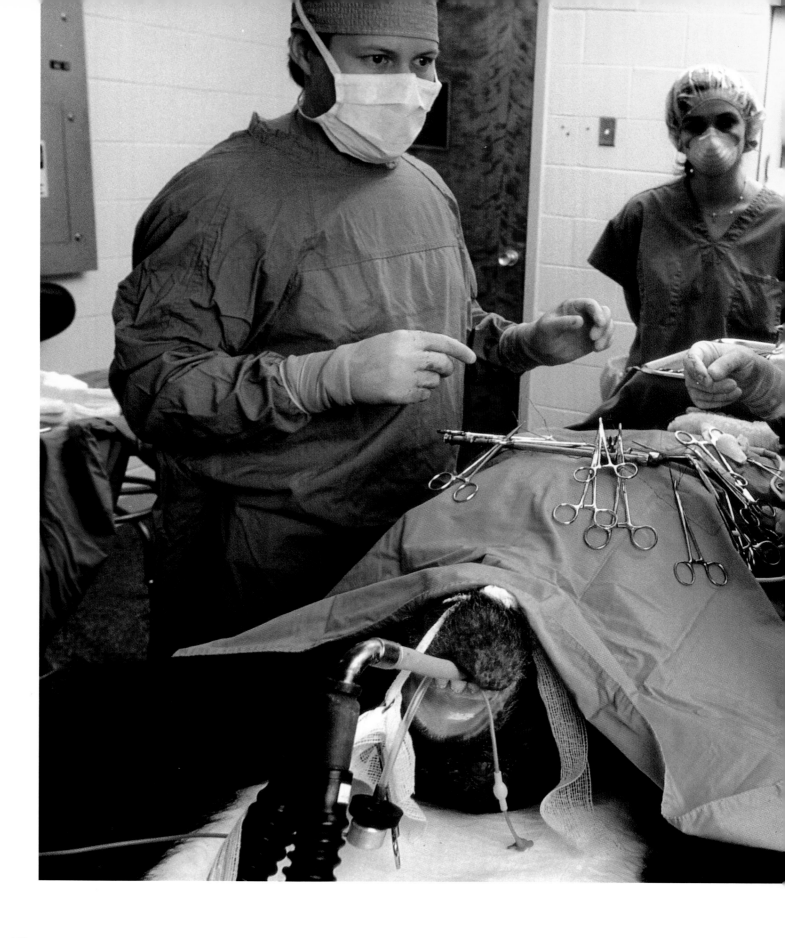

For experiments on hepatitis, a liver disease, veterinarians remove a section of chimp's liver at the Southwest Foundation for Biomedical Research in San Antonio, Texas. The donor chimp, under anesthesia, retained enough of this vital organ to survive. Biopsy has replaced sacrifice—standard practice in such studies thirty years ago. Below: At LEMSIP (Laboratory for Experimental Medicine and Surgery in Primates, an affiliate of the New York University School of Medicine), blood is drawn from a chimp under sedation and under restraint. HIV-infected white cells, separated from the blood, are sent to another laboratory for viral analysis. Stains between the arms are Betadine, not blood. Betadine sterilizes the skin before the needle goes in.

Quality nursery time finds Digger
and Gordo and Arden engaged
in affectionate babble with Cynthia
Kirby and chief veterinarian
James Mahoney—LEMSIP's soul
of compassion for research chimps.
Arden, eye to eye, has unmasked
the doctor; masks shield animals
vulnerable to human infections.
Pete Escobar holds a suckling
infant (above) in a nursery at
Southwest, where the chimp colony
numbers about 240. Here, too,
the lab provides enrichment,
including a retirement plan
for the research subjects.

Solitary or cramped confinement
at many primate labs has aroused
protest, not all of it lawful.
A masked reenactor displays
a photograph taken in 1986
when activists raided a Maryland
lab funded by the National
Institutes of Health. This facility
now has day rooms, night rooms,
and special play space in its 12
"biocontainment suites" for
infected young chimpanzees.

Research also confines
humans. Francine Patterson and
Ron Cohn cannot go off together
from the Woodside, California,
home where lowland gorillas Koko
and Michael have learned 800
words in American Sign Language.
A computer now gives Koko a "voice";
when she touches a screen icon,
the corresponding word is sounded.
In her enclosure the researchers
appear with Koko images— a stuffed
doll (at left) and a cutout.

Kanzi, star pupil of the Language Research Center near Atlanta, flaunts the record of a quiz on comprehension and syntax, using animal toys. Without gestural cues, a voice asked, "Can you make the doggie bite the snake?" The ten-year-old bonobo had never heard the sentence before, but he put the snake into the dog's mouth. Of 660 questions, Kanzi scored a correct response on 475 while a child of two was right on 435. Quick study, self-starter, whiz kid—the aura of prodigy surrounded Kanzi's early years. During his mother's lexigram lessons, he hung about, seemingly indifferent, sometimes bratty; yet he picked up the meanings of 150 symbols without training. His progress since then, under the tutelage of Sue Savage-Rumbaugh and her team, has rekindled interest in the contentious field of ape-language study.

FOLLOWING PAGES: All work and no play is not Kanzi's way. The distinguished lexigrammarian signaled "You chase" on the board held by Dr. Savage-Rumbaugh, then turned to his young friends, ready to run.

*Panbanisha finds the alphabet
primer an engrossing page turner;
as she ponders the pictures at
the Language Research Center,
a caretaker describes them.
Rare in the wild, bonobos, such as
Panbanisha, were little studied until
recent decades. Left: Panpanzee,
a common chimp, strolls off after
indicating—via lexigram—her
desire to visit one of 17 snacking
spots in the 55-acre preserve. Using
computers and speech synthesizers,
the LRC also teaches language arts
to nonspeaking retarded children.*

B ottle feeding followed by a
kiss, Linus the baby orangutan basks
in the arms of Rosanne D'Ercole at
their home in Ramsey, New Jersey.
Linus shares the affection and the
house with half a dozen other apes.
Their mistress says her husband owns
a good business so her animals
need not earn their keep. Instead,
they visit schools to enlist support for
endangered species, or bedsides to
liven a day for terminally ill children.
Primate mementos fill the D'Ercole
home—Bonzo with future President
Ronald Reagan in a prime spot.
Where's Linus? Somewhere
among the stuffed playmates that
fill his playpen (below).

PRECEDING PAGES: A former carnival
chimp, abandoned by the show,
Sam spends his years caged beside
his owner's bar in Maineville,
Ohio. In spite of posted signs
visitors still bootleg smokes to him;
he lights up for himself.

Cream or lemon, Coby? Vanessa and the household chimp compose the teatime scene as the girl's mother, Gini Valbuena, records it. Valbuena is a professional photographer in Clearwater, Florida. She owned apes purely for the pleasure of their company, she says, until state law banned the keeping of such pets. The law classes the great apes with the great cats and other large, dangerous species. Florida requires a commercial use permit, allowing experienced handlers to keep apes for exhibiting, sale, or other commercial purposes. So Coby models, though his image rarely sells. Only one sold in 1992, for $400. With expensive cages indoors and out, a hearty appetite, vitamins, toys, videos for a tube addict, the sale "didn't begin to cover expenses," says Valbuena, "but it keeps me legal." Coby has reached sexual maturity, and among strangers he displays a chimp's natural urge to dominate; he doesn't go out much. When guests visit and Coby acts up, his bête noire—King Kong on video—drives him into hiding under a blanket.

Mr. Jiggs revs up a lively party
in New Jersey, with trainer Ron
Winters as copilot. "World's
smartest chimp" also does TV
shows, cocktail parties, bar
mitzvahs. She—forget the stage
name—shoots pictures and
pistols, mixes drinks, hugs guests.
"Fully toilet trained," says her
flyer. Her front teeth are gone;
she performs with jaws tied shut.
Above, she holds a portrait
of herself at 18 months, fresh
out of Africa. In 1991 she
and Winters moved to California;
they now live in Encinitas.

Florida retirees: Mae and Bob Noell settled down at their Chimp Farm in Tarpon Springs, after 31 years of trouping a "Gorilla Show" through small towns of the eastern seaboard. "We retired our animals with us," she wrote. "They had earned their living as well as we had." For this portrait, a poster calls up memories. The Noells' routine let a challenger sign a release, put on a football helmet, and earn five dollars trying to best an ape that wore a muzzle and leather mittens. The ape was always a chimpanzee, but patrons in the '40s found that word too hard—they turned it into "chimpaneze," "chipmonk," or just "varmint," but they could all say "gorilla." More than 35,000 humans, including a 250-pound pro wrestler, lost to a 95-pound chimp. No human ever won, but a few drunks reportedly sobered up for good. In 1971 the Noells took the show off the road. Their "farm" became a breeding colony and a refuge for unwanted or aging primates. Snookie and Joe, stars of the '50s, passed on. Bob Noell died in 1991. Mrs. Noell remains, with old ape friends and new ones. Some are still fruitful, and multiply.

California retirees: Cheeta, hero's helper in several Tarzan movies, lived with his trainer Tony Gentry (below)—and visited co-star Johnny Weissmuller (left) before the actor died in 1984. Nearing 60, the chimp joined the ménage of trainer Dan Westfall in Palm Springs and took up abstract painting. Says Westfall, "I think there are a lot of animal lovers who would like an original Cheeta."

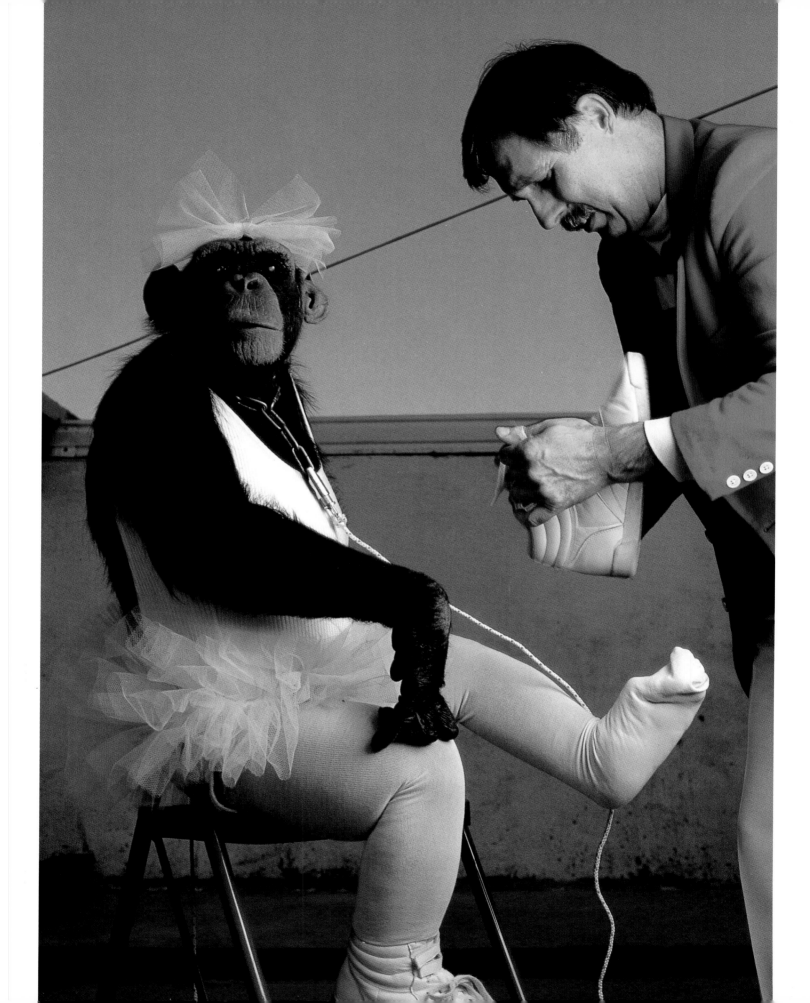

In ballerina finery or facsimile jungle, chimps play a diversity of roles. Dan Westfall shoes Susie, one of the Marquis Chimps, a unicyclist on the Ed Sullivan Show. The old TV favorite is history, but this forty-something ingenue still works nightclubs, in tux or tutu. She helped train her trainer, a neophyte when he bought her. "If I did something wrong, Susie straightened me out," says Westfall. Above the costume frills dangles a chain, reminder of the animal's formidable strength. Christopher (above) roams man-made wilds at Jungle Larry's Zoological Park in Naples, Florida. Though he never worked with elephants, he posed with baby Hadari at a trainer's bidding. His home is an island, one of several stocked with wildlife; tourists see them from boats. The park has shifted its focus from sheer entertainment to educational themes.

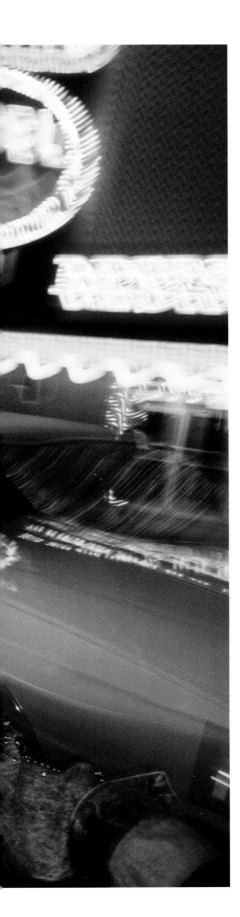

"Glitter Gulch"—more formally,
Fremont Street in downtown
Las Vegas—gives Bo a gaudy route
for an evening ride with her
owner, veteran animal trainer
Bobby Berosini. She loves
to ride in the car, he says.
Above, Bo appears in rock star
attire to shake hands with
spectators after her set in
"Lido de Paris." Berosini says
his act is "about a man trying
to be in charge of 5 orangutans
who in turn just keep making
a monkey out of him" and
appeals to people who would
enjoy telling off their bosses.

Music critic in mime, Tiga covers her ears as Berosini sings—and then closes his mouth. Rusty (at left), a young male, stands by as Bo plays for a laugh with a gesture considered lewd by Americans but meaningless in itself to orangutans. After 16 years in Las Vegas, the Berosini show moved in 1993 to a new country-music center: Branson, Missouri.

FOLLOWING PAGES: Exit the stand-up comic, enter the heavy, miscast as usual. King Kong aims a glare at tourists in Universal Studios Hollywood. Contrary to movie make-believe, nature made gorillas lotus-eaters—not man-eaters.

Peter Elliott, great at great ape
imposture, suits up for his gorilla
part in Eugene O'Neill's drama
The Hairy Ape. *Close study of
his role models—including the
University of Oklahoma's famous
chimpanzee colony—gave the
English actor a keen understanding
of primate gaits and gestures.
Arm extensions permit him the
sloping stance of a knuckle-walker.
Elliott shared the ape lead in the
filming of Dian Fossey's* Gorillas
in the Mist; *he played some close-up
scenes as Digit, as well as the adult
female Simba. Humans aping apes,
apes using new devices of human
language— role reversals as striking
as this suggest new reasons to hope
the great apes will survive. Elliott's
résumé says, however, that his
performance in costume once
frightened policemen into trying
to shoot him out of a tree
in a city park. Ironies abound
in zones where the worlds of men
and apes meet and overlap.*

"Eerie souls in animal furs"—Adriaan Kortlandt's phrase for chimpanzees he studied in 1960 in the Belgian Congo, and this portrait of a pensive adult bonobo, evoke the paradoxical qualities of the great apes. Even as we humans begin to understand them, their beleaguered status in the wild poses a challenge to our foresight, our imagination, and our ability to share the world with others of our kind.

By George B. Schaller

An Epilogue

Zoologist R. L. Garner went into the forests of West Africa during the 1890s to conduct the first study of chimpanzees and gorillas in the wild. There he assembled a steel-mesh and iron cage for himself and sat in it for over one hundred unproductive days. The cage was to protect him from dangerous animals—if need be, from what one horrified writer called the "vile propensities" of the chimpanzee and especially from that "frightful production of nature," as an 18th-century sea captain described the gorilla.

Over six decades later, in 1959, I went with my wife, Kay, to observe mountain gorillas for more than a year in the Virunga volcanoes of the Belgian Congo. Daily I followed fresh gorilla trails through thick undergrowth until the apes and I met in peace, not afraid but curious about each other. At times I wondered if they recognized the ancient connections that bound us.

Garner's and my approach to a field study differed somewhat, but we both saw in the apes a primal part of humanity's heritage. They raised the fundamental question, Who am I?

After Charles Darwin published *On the Origin of Species* in 1859, humankind was brusquely pushed from the center of the universe to jostle with the apes. Many eventually accepted this relationship but became bemused, confused, and uncertain because human values had suddenly become relative. As the wife of the Bishop of Worcester reportedly exclaimed when she heard of Darwin's ideas: "Descended from the apes! My dear, let us hope that it is not true, but if it is, let us pray that it will not become generally known." At least, it was thought in those years, some human races had descended farther from the apes than others. And, of course, in European opinion the Europeans had left the apes farthest behind.

With anthropological condescension, scientists placed humans in god-like isolation into the family Hominidae and relegated the four great apes into the family Pongidae. We have sought to maintain our perceived uniqueness, to draw a sharp line between man and beast: man the political animal, the tool-making animal, the rational animal, the speaking animal. Humans are, of course, physically distinctive, not least for the large brain that has led to a certain mental agility, semantic skill, and intelligence, often misapplied.

But to be consciously human, all of us must "sweep away this vanity," in Thomas Henry Huxley's words, and look upon the apes without prejudice. According to recent and much-debated molecular studies, the human, the chimpanzee, and the gorilla lineages diverged sometime between six and ten million years ago; the orangutan is a more distant relative. The apes still live within us; they are not living fossils but living relics of our earliest ancestors, closer to us in body and mind than any other creatures. Humankind cannot define itself without reference to them.

Like ours, ape society is rich in tribal traditions. Regional dialects exist: The pant-hoot calls of male chimpanzees at Gombe differ from those at Mahale. Moreover, anyone who studies apes soon realizes that he or she is observing individuals. Each animal has its own temperament, mental prowess, motives, aspirations, and desires. Each affects its own society by its idiosyn-crasies, friendships, animosities, and kinship ties.

We should, of course, study apes just for themselves, not as templates of human society. Yet we cannot help but judge them by human standards because they are so like us. Any parallels are immediately grasped and too often carelessly used to infer kinship. Yet each ape species has its own evolutionary history, its society shaped by social and ecological pressures different from our own.

Apes do not live in a zoological vacuum. Humanlike attributes are found not only in primates but in other animals as well. Many creatures use tools, among them digger wasps and sea otters. Various birds learn regional song dialects. Deadly encounters, infanticide, and cannibalism have been reported in many mammal species, including chimpanzees. Once, years ago, I studied lions in Tanzania's Serengeti National Park. I observed that male lions in a pride may form a coalition and, in effect, make war on the males in a neighboring pride in an attempt to gain access to more females. At Gombe in the 1970s, the males of one chimpanzee group defeated and killed off the males of another. Chimpanzees of either sex may help an ill or injured group member, but elephants may also do this.

Nevertheless, as our close kin, apes are windows to our minds. They provide experimental models useful in highlighting issues on the evolution of intelligence. They also provide hypotheses to test the development of our one truly distinctive innovation: symbolic language. Apes cannot articulate the sounds of human speech, but they do have some ability to learn a human sign language. They can communicate simple thoughts and desires, and they can refer to something not present, and ask questions. But so far they have not revealed an ability to learn the complicated syntax that human children master so easily.

We tend to forget that animals need not be able to emulate us to be highly intelligent. Each species has its own perspective on the world. Scientists all too often have had a penchant for teaching apes what they cannot do well rather than measuring abilities at which apes excel. Apes did not evolve to make abstract signals with their fingers or sit alone in a laboratory to be tested with bits of plastic of different color and shape. They evolved an intelligence for solving social problems. They have an immensely rich emotional life. They are intensely aware of their social world to which they constantly adjust with insight as they form alliances, appease, compete, reconcile, cooperate, plan. Apes understand, it seems, each other's goals and motives; they are self-aware, consciously thinking about objects and events. They deceive. One chimpanzee limped only when in sight of the chimpanzee that caused its injury. Yet so far there is little evidence for empathy—a chimpanzee may show mental anguish, but does it project that feeling into another? Is altruism important in ape society? Like the hero in Matthew Arnold's poem, apes seem to wander "between two worlds, one dead, / The other powerless to be born." I wonder if a captive gorilla dreams about the forests of its childhood.

I do not wish to suggest that we should study apes because it is useful to do so. Above all, we should aspire to their company because it gives us pleasure, because we find them elegant and impressive with societies that are fascinating. There is intimate beauty when we are in close contact with them, accepted by them as peaceful aliens in their forest home. I know that the apes are my kin, but when looking into their eyes I can feel it with all my being.

I believe that when studying apes, or indeed any animal, one must consider not just science but also philosophy and ethics. Scientists have a moral obligation to help protect and preserve the species they study. I am purposefully mixing science and values: Apes are our sibling species to whom we owe respect, empathy, moral reflection, and certain rights—above all, the right of survival.

Of about 170 primate species on earth, most live in tropical forests. About half are already at great risk by having their habitat destroyed, being hunted for food, and being exported for pets and biomedical research. In Africa there are perhaps 200,000 chimpanzees, at most 20,000 bonobos, and probably no more than 75,000 gorillas (a mere 600 belonging to the mountain subspecies). About 20,000 orangutans are thought to survive in Sumatra and Borneo. By contrast there are more than five billion people, increasing at the current rate of ninety million a year. The apes are in decline, and the remaining populations are becoming fragmented as their forests are cleared at an ever increasing rate. Hunted, affected by inbreeding, their societies disrupted, small populations will ultimately vanish. Even now many are imperiled by political upheaval and civil wars. As Jane Goodall noted in her foreword to the book *Understanding Chimpanzees,* if present trends continue "in another 100 years there won't be any more chimpanzees."

Like all organisms, the apes have become victims of human excess and our barren zeal to conquer the environment. We must protect the apes in large forest tracts not just on the basis of moralistic sentiment but as the means to a larger goal: to save with compassion the remnants of nature and the diversity of life. In this the apes can assist us, for they serve as powerful symbols of diverse and valuable ecosystems with all their plant and animal species. Humankind must develop new attitudes toward the natural world and change its habits, appetites, and expectations. We must foster a respectful relationship with the land, a willingness to share this planet with other forms of life. The apes impinge uniquely on our consciousness, emphasizing by their very existence that a species, any species, is worth preserving— that all testify to the wonders of creation.

Nevertheless, the apes have been and are still being denigrated and persecuted much as indigenous peoples were treated by conquerors in the past. By viewing such peoples—whether Amerindians, Australian Aborigines,

Hottentots of southern Africa, or others—as uncivilized and savage wanderers of the wilderness, incapable of full humanity, colonists once felt entirely justified in killing them and taking their land. For example, Robert Gray wrote in 1609, in his little book *A Good Speed to Virginia,* much of the world is "possessed and wrongfully usurped by wild beasts...or by brutish savages, which by reason of their godless ignorance, and blasphemous idolatry, are worse than those beasts...."

In 1906, the Bronx Zoo briefly exhibited a Pygmy—a man named Ota Benga, from what is now Zaire—in the same cage of the monkey house as a young orangutan. Commenting on this spectacle the *New York Times* said that whether the Pygmies "are held to be illustrations of arrested development, and really closer to the anthropoid apes than the other African savages, or whether they are viewed as the degenerate descendants of ordinary negroes, they are of equal interest to the student of ethnology, and can be studied with profit." At least the editorial writer also noted that the show was "not exactly a pleasant one," serving only to gratify "an idle curiosity and a rather brutal sense of humor."

Racism taken for granted in the past is discredited now. Certainly no scientist would speak of any human group as "closer to the anthropoid apes" than any other. Morality toward indigenous peoples has fortunately changed. Instead of being viewed as uncivilized brutes, such peoples are now admired for their ecological knowledge and distinctive cultures; they are treasured as examples of human diversity. The slave trade, which in the past century created such moral outrage, was stopped by international action. The same generosity of spirit must now be extended to the apes. We must cease the wanton killing and the destruction of their forest homes, and we must cease treating them as inferior creatures to be captured and sold. Garner put it well: "to dignify the apes is not to degrade man but rather to exalt him."

In his 1863 classic *Man's Place in Nature,* Thomas Henry Huxley wrote that "thoughtful men, once escaped from the blinding influences of traditional prejudice, will find in the lowly stock whence man has sprung, the best evidence of the splendor of his capacities; and will discern in his long progress through the Past, a reasonable ground of faith in his attainment of a nobler Future."

If we lose the apes we will have deprived ourselves of a nobler future. We will become isolated and meaningless fragments, half beast and half human, without an evolutionary bridge to connect the past to the future.

The millennium is almost here. It can be an end or a beginning. We must look into ourselves and find what it means to be human. Our intellect, passions, hopes, dreams, and spiritual splendor must all be used to preserve the world's natural heritage, its biological treasures, and with it the great apes, those living monuments to our ghostly past. If we do this, it will be a new beginning for humankind.

At home in the Taï Forest, a young chimp must apply its intelligence to many things: danger, from leopards; food, from many plants and several species of smaller mammals; and skills, social as well as material. Today success depends on learning, for apes and humans alike. How the earliest humans succeeded may yet become clearer as researchers follow the apes in the wild—much as Louis Leakey expected. In any case, writes George Schaller, the living apes emphasize "by their very existence that a species, any species, is worth preserving—that all testify to the wonders of creation."

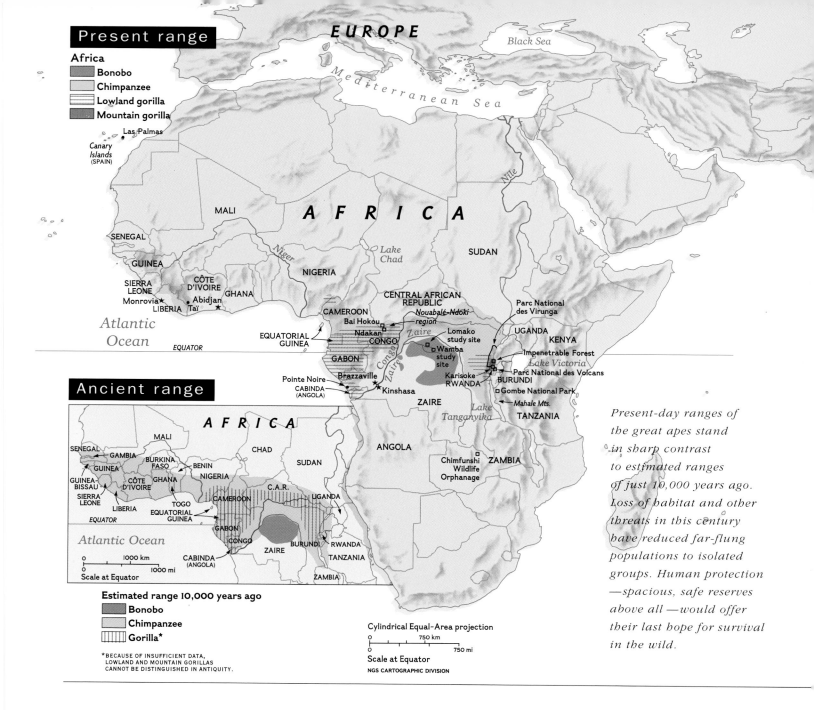

EUROPE

Black Sea

Mediterranean Sea

Las Palmas

Canary Islands (SPAIN)

AFRICA

MALI

SENEGAL

GUINEA

SIERRA LEONE

CÔTE D'IVOIRE

GHANA

Monrovia★

LIBERIA Taï★

Abidjan★

Atlantic Ocean

EQUATOR

Niger

Lake Chad

NIGERIA

SUDAN

Nile

CAMEROON

EQUATORIAL GUINEA

GABON

Bai Hokou

Ndakan

Pointe Noire

Brazzaville

CABINDA (ANGOLA)

Kinshasa

CENTRAL AFRICAN REPUBLIC

Nouabalé-Ndoki region

CONGO

Zaire

Congo

Zaire

Lomako study site

Wamba study site

Karisoke

RWANDA

BURUNDI

ZAIRE

Parc National des Virunga

UGANDA

KENYA

Impenetrable Forest

Parc National des Volcans

Lake Victoria

Gombe National Park

Mahale Mts.

Lake Tanganyika

TANZANIA

ANGOLA

Chimfunshi Wildlife Orphanage

ZAMBIA

AFRICA

MALI

SENEGAL

GAMBIA

GUINEA

GUINEA-BISSAU

SIERRA LEONE

LIBERIA

CÔTE D'IVOIRE

BURKINA FASO

BENIN

GHANA

TOGO

NIGERIA

CHAD

SUDAN

C.A.R.

CAMEROON

EQUATORIAL GUINEA

GABON

CONGO

CABINDA (ANGOLA)

ZAIRE

BURUNDI

RWANDA

UGANDA

TANZANIA

ZAMBIA

EQUATOR

Atlantic Ocean

0   1000 km
0   1000 mi
Scale at Equator

Cylindrical Equal-Area projection

0   750 km
0   750 mi

Scale at Equator

NGS CARTOGRAPHIC DIVISION

Present-day ranges of the great apes stand in sharp contrast to estimated ranges of just 10,000 years ago. Loss of habitat and other threats in this century have reduced far-flung populations to isolated groups. Human protection —spacious, safe reserves above all —would offer their last hope for survival in the wild.

Those wishing to support projects for research, care, and conservation of the great apes may note the following addresses.

African Wildlife Foundation
1717 Massachusetts Avenue, N.W.
Washington, D.C. 20036, U.S.A.

Chimfunshi Wildlife Orphanage
P.O. Box 11190
Chingola, Zambia
or
c/o The Jane Goodall Institute

Committee for Conservation and Care of Chimpanzees
3819 48th St., N.W.
Washington, D.C. 20016, U.S.A.

The Dian Fossey Gorilla Fund
45 Inverness Drive East, Suite B
Englewood, Colorado 80112-5480, U.S.A.
or
110 Gloucester
Primrose Hill
London NW1 8JA, U.K.

The Gorilla Foundation
Box 620-530
Woodside, California 94062, U.S.A.

The Jane Goodall Institute
P.O. Box 41720
Tucson, Arizona 85717, U.S.A.
or
15 Clarendon Park
Lymington, Hants. SO41 8AX, U.K.

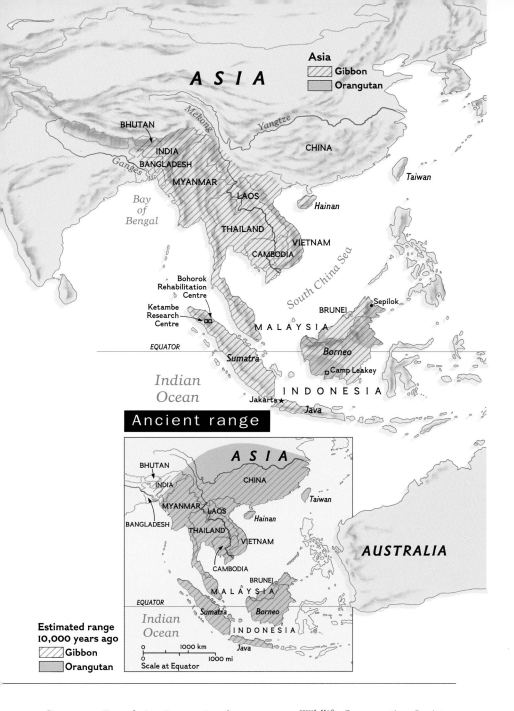

Asia
- ▨ Gibbon
- ▩ Orangutan

ASIA

BHUTAN
INDIA
BANGLADESH
MYANMAR
CHINA
Mekong
Yangtze
Taiwan
Hainan
LAOS
THAILAND
VIETNAM
CAMBODIA
Bay of Bengal
Ganges

Bohorok Rehabilitation Centre
Ketambe Research Centre

South China Sea
BRUNEI
Sepilok
MALAYSIA

EQUATOR
Sumatra
Borneo
Camp Leakey

Indian Ocean
INDONESIA
Jakarta ★
Java

Ancient range

ASIA
BHUTAN
INDIA
CHINA
Taiwan
MYANMAR
LAOS
Hainan
BANGLADESH
THAILAND
VIETNAM
CAMBODIA
BRUNEI
MALAYSIA
EQUATOR
AUSTRALIA
Sumatra
Borneo
Indian Ocean
INDONESIA
Java

Estimated range 10,000 years ago
- ▨ Gibbon
- ▩ Orangutan

0   1000 km
0   1000 mi
Scale at Equator

Orangutan Foundation International
822 S. Wellesley Avenue
Los Angeles, California 90049, U.S.A.

Primarily Primates, Inc.
P.O. Box 15306
San Antonio, Texas 78212-8506, U.S.A.

Projet de Protection des Gorilles
Howletts & Port Lympne Foundation
Brazzaville BP 13977
Republique du Congo
or
c/o John Aspinall's Wildlife Sanctuaries
750 Lausanne Road
Los Angeles, California 90077, U.S.A.

Wildlife Conservation Society
The Wildlife Conservation Park
Bronx, New York 10460, U.S.A.

World Wide Fund for Nature
(WWF Switzerland)
Forrlibuckstrasse 66
Postfach, 8037
Zurich, Switzerland

Acknowledgments

The Book Division gratefully acknowledges the assistance of all who have helped make possible the publication of this volume—in particular, the individuals and organizations named, portrayed, or quoted in its pages. It thanks government officials and personnel of the nations concerned, especially those actively defending ape populations; members of the scientific community; and members of the public whose support is crucial to research and conservation.

Michael Nichols would like to thank all those who helped him in his photographic quest, but especially Geza Teleki, who inspired him to expand his work to the scope of investigation that is presented here.

The division extends special thanks to those cited here: Shizuko Aizeka, Susan Allen, Dean Anderson, Susanne Anderson, Mark Attwater, Robert C. Bailey, Bill Beers, Catherine Belden, Allard Blom, Christopher Boehm, Monica Borner, Lisa Brock, Betsy Brotman, Thomas M. Butler, Robert L. Carneiro, Carol Lee Carson, Jon Charles Coe, Margaret Cook, Gillian Dundas, Robert J. Edison, Ruth Eichhorn, Kathy Espin, John G. Fleagle, Roger Fouts, Mary Geib, Vanne Goodall, Kenneth Harris, Matthew Hatchwell, Beth Heng, Michael Hutchins, Karen S. Killmar, Ruth M. Keesling, Betty Ktana, John C. Landon, Autumn Marie Latimore, Rick Lee, Carol Leifer, Nicholas Leon, Rodney MacAlister, Patricia M. McGrath, William C. McGrew, Stacy Maloney, Terry L. Maple, Linda A. Marquardt, Valerie Mattingley, David Maybury-Lewis, Aly Mjenga, Godefroid Manirankunda, Jan Moor-Jankowski, Krishna K. Murthy, Michael Nee, Carole Noon, John Oates, Alex Peal, Randall Lee Pouwels, Wendy Pratt, Alfred Prince, Rebecca Rooney, Sandy Rowland, Brooke Russell, Shawn Sandor, Patricia Sass, Craig Sholley, Lynnette Simon, Alan Sironen, Susan Stenquist, Laura Strickland, William C. Sturtevant, Jito Sugardjito, Diane Sunderland, Mary Beth Sweetland, Wallace W. Swett, David Tetzlaff, Nancy Tetzlaff, Jerry Thompson, Nancy Thompson-Handler, Robert M. Utley, Jeanette Wallace, Dan Wharton, Natasha Yankoffski, and Jim Yeager.

Index

Boldface indicates illustrations; *italic* refers to picture captions. For ease of identification, main entries that are names of apes are given in blue.

MICHAEL NICHOLS BY PETER WILKINS

JANE GOODALL BY MICHAEL NICHOLS

MICHAEL NICHOLS He grew up in Muscle Shoals, Alabama, with a collection of GEOGRAPHICS. The U.S. Army trained him in photography in the early 1970s; then he studied photography and fine arts at the University of North Alabama, graduating in 1977. His high-risk, award-winning work—in deep caves, on an Arctic cliff, down whitewater rivers—inspired his nickname "the Indiana Jones of Magnum [the famous photo agency]."

JANE GOODALL As a child in Bournemouth, England, she wrote descriptions of the habits of mammals and birds around home. A secretary at 20, she saved her money for a visit to Africa. Her study of wild chimpanzees in Tanzania began in 1960. She received a Ph.D. from Cambridge University in 1965; her monograph, *The Chimpanzees of Gombe,* appeared in 1986. Thereafter she began her crusade for conservation of wild chimpanzees and better treatment of captives.

GEORGE B. AND KAY SCHALLER COURTESY GEORGE B. SCHALLER

MARY G. SMITH BY DEVRA G. KLEIMAN

GEORGE B. SCHALLER Born in Germany, now a U.S. citizen, he earned B.A. and B.S. degrees at the University of Alaska, a Ph.D. at the University of Wisconsin in 1962. His affiliation with the New York Zoological Society began in 1966. As a field biologist, he has worked in Alaska, Zaire, India, Tanzania, Pakistan, China, Mongolia, and other countries—usually with his wife, Kay. His publications include some 60 technical and as many popular articles (5 in the GEOGRAPHIC), 2 children's books, 5 monographs, and 5 popular books; the latest is *The Last Panda.*

MARY G. SMITH She grew up in a military family and without a hometown, finishing her schooling in Paris and London. In 1956 she joined the NATIONAL GEOGRAPHIC staff as a picture editor. For more than 30 years she has worked with scientists who receive grants from the Society's Committee for Research and Exploration, helping them report their projects to a popular audience through magazine articles, books, or TV. Widely traveled, she is a board member of five organizations dedicated to research and conservation.

International Editions

In addition to English, *The Great Apes: Between Two Worlds* is co-published in the following languages: German, Japanese, and Spanish, under the guidance of the International Publications program, Alice J. Dunn, Carolyn L. Goble, Anne E. Wain, coordinators.

Composition for this book by the National Geographic Society Book Division with the assistance of the Typographic section of National Geographic Production Services, Pre-Press Division. Printed and bound by R.R. Donnelley & Sons, Willard, Ohio. Color separations by Graphic Art Service, Inc., Nashville, Tenn.; Lincoln Graphics, Inc., Cherry Hill, N.J.; and Phototype Color Graphics, Pennsauken, N.J.

Library of Congress CIP Data
Nichols, Michael.
   The great apes : between two worlds / photographs and essays by Michael Nichols ; contributions by Jane Goodall, George B. Schaller, Mary G. Smith ; prepared by the Book Division, National Geographic Society.
      p. cm.
   ISBN 0-87044-947-8 (reg. edt.). — ISBN 0-87044-948-6 (deluxe edt.)
   1. Apes. 2. Chimpanzees. 3. Gorilla. 4. Orangutan.
I. National Geographic Society (U.S.). Book Division. II. Title.
QL737.P96N54 1993
599.88'4—dc20   93-2261
                                   CIP